and
Stephanie L. Rhee

Multivariate Analysis of Categorical Data: Applications

Advanced Quantitative Techniques in the Social Sciences

VOLUMES IN THE SERIES

Multivariate Analysis of Categorical Data: Applications

John P. Van de Geer

Advanced Quantitative Techniques
in the Social Sciences Series **3**

SAGE Publications
International Educational and Professional Publisher
Newbury Park   London   New Delhi

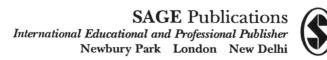

Copyright © 1993 by Sage Publications, Inc.

Previously published by Van Loghum Slaterus, copyright 1988, as *Analyse van Kategorische Gegevens*. Translated by the author.

For information address:

 SAGE Publications, Inc.
2455 Teller Road
Newbury Park, California 91320

SAGE Publications Ltd.
6 Bonhill Street
London EC2A 4PU
United Kingdom

SAGE Publications India Pvt. Ltd.
M-32 Market
Greater Kailash I
New Delhi 110 048 India

Printed in the United States of America

Library of Congress Cataloging-in-Publication Data

Geer, J. P. Van de (John P.)
    Multivariate analysis of categorical data / John P. Van de Geer.
        p.   cm.—(Advanced quantitative techniques in the social
    sciences; 2-3)
    Includes bibliographical references and indexes.
    Contents: v. 1. Theory—v. 2. Applications.
    ISBN 0-8039-4565-5 (v. 1).—ISBN 0-8039-4564-7 (v. 2)
    1. Multivariate analysis.   I. Title.   II. Series.
QA278.G433   1993
510.5'35—dc20   93-3706

93   94   95   96   10   9   8   7   6   5   4   3   2   1

Sage Production Editor: Diane S. Foster

Contents

Series Editor's Introduction

In our introduction to Volume 1 of Van de Geer's book, we saw that there was a heavy emphasis on the geometry of Exploratory Multivariate Analysis. In Volume 2 there is an equally heavy emphasis on computing, and the computing is done with ready-made programs incorporated into *Categories,* a sub-package of *SPSS* (*SPSS Categories,* 1990). The emphasis on the geometry is still there, and there any many interesting pictures of data analysis results, but following GIFI (1990) the book is now structured in chapters corresponding with specific computer programs.

Any of the programs discussed is a statistical technique in the sense that one inputs data from a set of possible input data structures, and the program then generates results, which are supposedly a unique function of the input. A complicated function, to be sure, because there is numerical output, there are graphs, and there even may be error messages. It is still useful to think of a program (or a package, which consists of a number of programs) as something that transforms input to output, where the precise nature of the function implemented in the program is often determined by some additional parameters defined separately from the input data. Van de Geer discusses some of the key GIFI programs, and show what they do to input data of various kinds.

Packages, such as *SPSS,* are not very popular with statisticians. They are associated with a rigid menu of canned statistical techniques, which are applied in a routine and mechanical way by people who really don't know what they are doing. In this respect they are similar to fast food or Hollywood movies. Convenient, yes, popular, yes, but superficial and without quality. I think the disdain of statisticians for package-statistics is rather elitist and short sighted. Statisticians are under the spell of mathematical notions of optimality, of philosophical notions about inference, and of scientific notions about the proper treatment of empirical data. The packages, however, provide a language to talk about data, which is understood by many scientists (and by many journal editors). Maintaining that packages are not a good language is quite silly. The alternative languages that statisticians have come up with so far, are either ridiculously impractical (decision theory is a good example) or they can only be applied to tiny examples or in well-controlled situations. Most applied statisticians, whether they are practicing Bayesians or not, simply use the packages on their clients' data sets. And many applied statisticians have found out, to their dismay, that there are a lot of things in data manipulation that can best be done in those packages.

Badmouthing the packages is like saying that if you want to talk about food, you can only do so properly in French. Saying that packages are used by people who don't know what they are doing is dangerously close to the sort of arrogance that mathematicians seem to be especially prone to. The people who are using the packages are not necessarily out there to discover rock-solid truths, they are there to publish papers in scientific journals in which the packages are the norm. Survey type data are often messy, they do not satisfy the standard models, and they burst out of the proper statistical frameworks. If the food we are talking about is just your daily run-of-the-mill food, we might as well talk about it in plain English.

Exploratory data analysis is another statistical bogeyman. Somewhere, somehow, statisticians got the idea that science (proper science, that is) proceeds in two steps. The first step is exploratory. The scientist does all kinds of dirty things to his or her data, things that are certainly not allowed by the canons of statistics, and at the end of this thoroughly unrespectable phase he or she comes up (miraculously) with a theory, model, or hypothesis. This hypothesis is then tested with the proper confirmatory statistical methods. Of course, Popper or no Popper, this is a complete travesty of what *actually* goes on in all sciences some of the time and in some sciences all of the time. There are no two phases that can be easily distinguished. There is no dirty and clean work, and for that matter the distinction between exploratory and confirmatory seems to allocate all the interesting and creative work to the exploratory phase anyway.

Enough of these generalities. The basic idea behind the GIFI system is that variables can be grouped into subsets in various ways, and variables can be quantified using various types of restrictions. By combining partitionings of the variables with general classes of measurement restrictions, we recover many of the classical multivariate techniques, but also many extensions. The chapters in Volume 2 are named after several of the more important techniques, with associated computer programs. *PRIMALS, HOMALS, ANACOR,* and *PRINCALS* are different generalizations of PCA; *CANALS* generalizes canonical correlation analysis; and *OVERALS* generalizes a form of multiple-set canonical correlation analysis. In a sense, all GIFI techniques are options in the *OVERALS* program, but special techniques require special algorithms and special input-output options, and thus special implementations. *HOMALS, PRINCALS, ANACOR,* and *OVERALS* have been incorporated in *SPPS Categories* and are consequently widely available. *ANACOR* is Correspondence Analysis, *HOMALS* is Multiple Correspondence Analysis, and consequently these two techniques are also available in *SAS* and *BMDP* under different names.

Another aspect that Van de Geer's book has in common with GIFI (1990) is that there are many real-life examples. Not too big, but not too small either. With real variables, often trying to answer some relevant policy-related question. The relevant part of the book by GIFI is called *The Proof of the Pudding*. Of course, you may not like the pudding. But all the ingredients are discussed quite clearly in these two volumes, and after digesting them, you will be prepared well to go deeper into the GIFI pudding, or to prepare one of your own.

JAN DE LEEUW
SERIES EDITOR

Preface

This work consists of two companion volumes, one with the subtitle *Theory* and the other with the subtitle *Applications*. These books have been written for researchers (and students) interested in nonlinear analysis of categorical variables, either because they want to perform such analysis themselves or because they want a better understanding of publications in which such analysis appears.

The term *nonlinear* sounds somewhat technical and awe-inspiring, perhaps. However, in the present books this term hardly plays a role. Instead, the emphasis is on analysis of data that contain categorical variables. Such variables sort objects into a limited number of distinct groups, called *categories*. A simple example is a questionnaire that asks respondents to mention their preference for one of the existing political parties. All we know from their answers is that respondents with the same answers are "similar" (because they belong to the same category) and that respondents with different answers are "dissimilar."

It is assumed that the reader is familiar with a number of basic statistical concepts, such as mean, variance, covariance, and correlation. In addition, it will be helpful if the reader has some knowledge about current techniques of multivariate analysis, such as principal components analysis and canonical analysis, although these techniques are explained in the text.

The explanations offered focus on the treatment of categorical variables, and thus are somewhat different from what is found in classical textbooks on multivariate analysis. Moreover, these explanations are based mainly on pictures, not on the underlying algebra. This approach may be somewhat frustrating for readers who are familiar with classical multivariate analysis, and who are used to explanations in terms of algebraic formulas. For such readers, in the present volumes (especially in the book on theory) some sections have been added that focus on algebra. These sections, the numbers of which are marked with asterisks, may be skipped by readers who are not primarily interested in algebraic background.

These books were not written for researchers from only one special scientific discipline. In fact, categorical variables occur in many different sciences, not only in social and behavioral sciences, but also in such fields as economics, linguistics, biology, market research, and political science. The volume on applications therefore makes use of examples from many different fields of study.

It is important to note that these books do not pretend to give a survey of all possible methods of handling categorical data. On the contrary, they are restricted to one particular class of such methods, sometimes called the *GIFI system*. The volume on applications discusses in some depth six GIFI methods that are currently available (included in an SPSS computer program package). The word *GIFI* refers to the collective nom de plume adopted by members of the Department of Data Theory, University of Leiden, for joint publications. For the present volumes I have received permission to draw freely from contributions of other members of the department; I would like to thank them collectively—there are too many names to list here. There are two exceptions, however, who should be named. The first one is Jan De Leeuw, now professor at the University of California, Los Angeles. Professor de Leeuw can be regarded as the founder of the system, the basis of which was put forward in his doctoral dissertation in 1973 (see De Leeuw, 1984a). The other exception is Mrs. Gerda Van den Berg, to whom I am indebted for assistance during the preparatory stage of the Dutch text. I would also like to thank Professor David Messick, University of California, Santa Barbara, for his willingness to check the correctness of this translation.

To conclude this preface, I would like to make a few technical remarks. First, the examples in the text, especially those in the volume on applications, serve only to illustrate what a user of GIFI methods may expect from them. It is not my intention to imply that a GIFI method is always the best possible one. It is quite possible that a GIFI method may not give a satisfactory answer to a question a researcher has in mind, and in such a case a better answer may be expected from a different method of analysis.

Second, in the volume on theory, the examples are artificial and relatively small, to make it possible for the reader to check the calculations with the use of a small pocket calculator. However, it should be noted that in the text the given numerical values often are rounded, and this may cause small discrepancies between results shown in the text and those a reader may find when checking the calculations.

Third, the two volumes contain cross-references to each other. For the sake of economy in printing, such cross-references have been abbreviated. For instance, if in the volume on theory one finds a reference to section A.3.1 (or to Figure A.3.1, or to Table A.3.1), it refers to Section 3.1 (or Figure 3.1, or Table 3.1) in the volume on applications. Likewise, in the applications volume, references to the theory volume begin with the letter T. If no additional A or T appears in a reference, it refers to the volume in which it is found.

JOHN P. VAN DE GEER

1 PRIMALS

1.1 Introduction

1.1.1 Derivation of PRIMALS

PRIMALS is an acronym made up of two parts. PRIM (from *prime*) refers to the objective of the analysis: to obtain a quantification of categories and of objects that is optimal in the *first* dimension. As noted in Section T.4.3, this is called *single quantification*. Once quantification has been obtained, the quantified variables are treated as if they are numerical.

The letters ALS are an abbreviation for *a*lternating *l*east *s*quares. This refers to the algorithm used to calculate the optimal quantification. This algorithm is described in Section T.4.7, where it is shown that the algorithm alternates between a calculation of object scores (as averages of the quantifications of those categories that apply to the object) and a calculation of category quantifications (as averages of scores of objects within the same category). Once such calculations converge, the solution is said to obey the principle of reciprocal averaging.

1

1.1.2 Algebraic and Geometric Details

For algebraic details, and for the rationale of PRIMALS, we refer again to Section T.4.3. The example given there, for single optimal quantification, is in fact an illustration of PRIMALS. Section T.4.3 also contains definitions of some technical concepts, such as MC (multiple coordinates), discrimination measure, and eigenvalue. Furthermore, the section shows how PRIMALS results can be visualized in graphs.

1.1.3 Basic Properties of PRIMALS

A point of departure is that PRIMALS is primarily interested in the first dimension for optimal quantification. The reason, loosely speaking, is that PRIMALS will be applied if it can be assumed that all variables measure about "the same thing," and that category quantifications therefore must be representative for what all variables have in common. It is the same reasoning as in applications of principal components analysis (PCA) to numerical variables. If all variables measure about the same thing, then they will be highly correlated with the first principal component, and it follows that this first principal component gives the best indication of what all variables have in common.

This first principal component itself is a weighted sum of the original variables. The weights are chosen in such a way that object scores on the principal component have large correlation with all original variables. More precisely, the weights are chosen so that the sum of the squares of these correlations is maximized. This sum of squared correlations is called the *first eigenvalue*.

Nevertheless, classical PCA treats all variables as numerical. This implies that in the analysis the a priori category quantifications are preserved. Given this restriction, the first eigenvalue is maximized. In PRIMALS, however (and in general in single quantification), this restriction is dropped. This means that the a priori quantifications may be replaced by new ones (called *optimal quantifications*), in such a way that the resulting eigenvalue will be larger than for any other quantification.

However, I would add two cautions. The first is that the reasoning above is strictly valid only if there are no missing data treated as passive. In this case, the output of the PRIMALS program gives discrimination measures that are the square of the correlations indicated above. But if there are passive missing data, this is no longer true (as explained in Section T.6.6). The second caution is rather trivial. Although it is noted above that the eigenvalue is the *sum* of the discrimination measures, PRIMALS output gives the eigenvalue as the *average* of them.

1.1.4 Examples

First, a researcher will in general be able to give some a priori quantification to the categories of a variable. After all, research is not a blind
process, and the researcher must have some idea about what the categories
stand for.

As a first example, take a questionnaire about political issues. One of
the items may contain the statement, "Differences in incomes should
become smaller." The respondent answers by selecting a response from a
scale with 7 categories, running from 1 = "I agree completely" to 7 = "I
disagree completely." This numbering corresponds with a political stand
running from "left" to "right." The questionnaire may contain many such
items, all of them ordered from left to right. But then the question arises
whether there could be a more adequate numbering than that running from
1 to 7. For instance, for some items it might be true that differences in
political stands of respondents express themselves mainly in the range
between 4 and 7, and that responses from 1 to 4 could as well be merged.
However, it is difficult for the researcher to make such a revision on an a
priori basis. Single optimal quantification then will reveal what the best
revision would be.

As a second example, consider a questionnaire item for which the
middle response is labeled "don't know." The question arises whether
respondents who select this category really are in the middle of the scale.
It might be that such respondents have extreme political stands on other
issues. In other words, the a priori quantification of categories, however
reasonable it seems to be on first sight, may turn out to be quite suboptimal,
and the quantification given by PRIMALS may indicate that there is a
better solution for quantification.

1.1.5 Variables Treated as Active or Passive

An example clarifies the issue here. Take a research project that includes
variables assumed to measure the same underlying dimension (such as
political stand, from left to right), but that also includes some other
variables. For instance, the other variables may give information about the
backgrounds of respondents (sex, age, county, profession, and so on). A
PRIMALS solution based on *all* variables in such a case could easily result
in an unwanted solution. The quantification given by PRIMALS might be
largely dominated by the background variables. The solution may even
depend upon trivial relations (such as that retired people are older, or that
respondents who own cattle are farmers and live in rural areas). The effect

will be that in such solutions the political variables are more or less ignored (in the sense that they obtain small discrimination measures).

It then makes sense to distinguish between the two types of variables, and to take the political variables as active and the background variables as passive. This means that the resulting category quantification will be based only on the (active) political variables, whereas the passive variables are ignored at this stage. In a subsequent stage, categories of the passive background variables still can be quantified, by taking the average score of objects within the same background category. In this way one can find out whether there are differences between male and female respondents, between age groups, and so on. And at this stage one also could calculate discrimination measures for the passive variables. The maximized eigenvalue, however, depends only on the discrimination measures of the active variables.

1.2 Structure of the PRIMALS Program

1.2.1 Single Optimal Quantification

The first block of the PRIMALS program gives single quantification of the categories of active variables. At this step the following results are included:

1. optimal category quantifications (standardized)
2. corresponding MC values of categories (defined as the average of the object scores within a category)
3. the discrimination measure of each active variable
4. the maximized eigenvalue (in the PRIMALS program defined as the average of the discrimination measures, not as their sum)

The computer program requires that categories of each variable are given an a priori quantification (in the format of natural numbers: 1, 2, 3, and so on). This a priori quantification may be entirely arbitrary, but it remains advisable to number the categories in a meaningful way. The reason is that the PRIMALS program may display, on request, transformation graphs, in which the optimal quantification is plotted against the a priori quantification. Such a graph has no meaning at all if the a priori quantification was chosen at random. But if the a priori quantification makes sense, the transformation graph could be interesting. It may show, for instance, that the transformation is monotonic, which implies that the variable is in fact treated as an ordinal variable.

If there are missing data that are treated as passive, definitions must be adapted as described in Section T.6.6. In this situation category quantifications, or object scores, no longer need to have zero mean. Also, discrimination measures no longer can be interpreted as squared component loadings.

On request, the PRIMALS program will also print the object scores, although very often there will be no need for them.

1.2.2 PCA Step

The next block of the PRIMALS program contains a PCA step. Assume for the moment that there are no missing data treated as passive. The PCA step then gives results for the first two PCA dimensions. These results are given twice: (a) based on the a priori quantification and (b) based on the single optimal quantification. This makes it possible for the reader to compare the two results. The user will find that the first eigenvalue after transformation is larger (or at least not smaller) than the first eigenvalue before transformation. On the other hand, optimal quantification is not concerned about the value of the second eigenvalue. So it may happen that the second eigenvalue before transformation is larger than the second eigenvalue after transformation.

One may wonder why this PRIMALS block looks only at two PCA solutions and not at more than two. The reason is simple. In applications of PRIMALS it is usually assumed that the active variables depend mainly upon only one underlying dimension. This assumption gets some support if after transformation the second eigenvalue is considerably lower than the first one. When two dimensions after transformation are given, the user can verify whether this assumption is true.

Suppose now that there are missing data and that they are treated as passive. We then meet some technical problems. A classical PCA solution (based upon a priori quantification of categories) assumes that there are no missing data. This implies that for the second PRIMALS block the user is obliged to assign numerical values to the missing data. In PRIMALS there are two options:

1. Missing data obtain quantification equal to zero (for the PCA both before and after transformation).
2. Missing data are quantified as the average score of those objects that have missing data on the same variable.

The first option has the effect that the first PCA solution for quantified variables becomes less good. The reason is simple. Optimal quantification just ignores the values that missing data might have. But setting these values equal to zero, and requiring that the PCA solution take those zero values seriously, implies that the optimal quantification becomes somewhat disturbed.

The second option, on the other hand, will have the effect that the PCA solution for the first eigenvalue becomes somewhat better than the PRIMALS solution, because the missing data are replaced by quantifications that are as much as possible in agreement with the optimal quantification.

PRIMALS will print, on request, two correlation matrices, the first one based on a priori quantification (before transformation) and the second based on optimal quantification (after transformation). For both matrices all eigenvalues will be shown, as well as component loadings of variables on the first two dimensions. The user should keep in mind, however, that these results (when there are missing data treated as passive) are influenced by the choice between the two options above, and that results for the correlation matrix after transformation may be somewhat different from those given in the first step of PRIMALS.

1.3 First PRIMALS Example: Crime and Fear

1.3.1 Introduction

The data for this example are taken from Cozijn and Van Dijk (1976). The example is restricted to only 14 of the variables used in their study. It is also restricted to objects without missing data, which results in $n = 1,216$ objects (this example is also discussed in GIFI, 1990).

The 14 variables are illustrated in Table 1.1. To summarize, there are (a) 6 variables related to opinions about crime prevention, (b) 4 variables related to general feelings of anxiety and fear, and (c) 4 background variables. Note that within the first group of variables, one might make a distinction between variables 1, 4, and 6 (related to "social prevention") and variables 2, 3, and 5 (related to "penal measures").

The basic question, at this moment, is to find out whether there is a common dimension underlying opinions about crime prevention and feelings of anxiety. This explains why the first 10 variables are taken as active, whereas the 4 background variables are treated as passive.

1.3.2 PRIMALS Solution

The PRIMALS solution for single quantification gives a first eigenvalue (average of discrimination measures) equal to .242. The value is relatively small, and indicates that the 10 active variables do not have much in common.

TABLE 1.1 Summary of Variables in Crime and Fear Example

Variables 1-6 ask for opinion about effectiveness of methods for crime prevention; five response categories. Response "very effective" is category 5 in variables 1, 4, and 6, and category 1 in variables 2, 3, and 5.

1. Reeducation
2. Locking up
3. More severe punishment
4. Social work
5. Labor camps
6. Better employment

Variables 7-10 contain statements about feelings of fear or anxiety. Response 1 = "agree," 2 = "don't know," 3 = "disagree."

7. Watch out when you walk in the city.
8. It is unwise to go outdoors at night.
9. You cannot rely on the police.
10. When something happens to you in the street you cannot expect help from bystanders.

Variables 11-14 ask about background.

11. Religion (1 = Calvinist, 2 = Protestant, 3 = Roman Catholic, 4 = other, 5 = none).
12. Voting behavior (1 = PvdA, 2 = VVD, 3 = KVP, 4 = AR, 5 = CHU, 6 = PPR, 7 = abstain, 8 = don't know, 9 = don't vote, 10 = other).
13. Occupational status and sex, responses 1 to 7 for males and 8 to 14 for females (1 = higher employee, 2 = middle employee, 3 = small business, 4 = lower employee, 5 = skilled labor, 6 = unskilled labor, 7 = no profession).
14. Age (1 = 16-17, 2 = 18-24, 3 = 25-34, 4 = 35-49, 5 = 50-64, 6 = 65-70).

Table 1.2 shows that discrimination measures are relatively high for the 6 prevention variables and low for the 4 fear variables. This indicates that the prevention variables do not measure the same thing as the fear variables. We come back to this result in Section 1.3.4.

Table 1.3 gives the optimal quantifications. It shows that these quantifications are close to monotonic transformations of the a priori quantification. For the 4 fear variables, the transformation is almost linear. For the 6 prevention variables the relation is slightly less simple.

TABLE 1.2 PRIMALS Discrimination Measures in Crime and Fear Example

1	.312	6	.365	13A	.008
2	.291	7	.085	13B	.026
3	.454	8	.122		
4	.453	9	.147		
5	.146	10	.041		

TABLE 1.3 PRIMALS MC Quantification in Crime and Fear Example

Variables	*1*	*2*	*3*	*4*	*5*
1	−.181	−.606	−.216	.309	.820
2	−.575	−.000	.746	.726	1.364
3	−.495	.324	.891	1.194	1.726
4	−1.656	−.842	−.787	.054	.835
5	−.407	.129	.016	.456	.803
6	−1.825	−.648	−.652	.070	.684
7	−.316	−.062	.310		
8	−.473	−.226	.291		
9	−.519	−.109	.364		
10	−.322	−.092	.160		

1.3.3 Passive Interactive Variable

Variable 13 is treated as a passive variable, but it is also an *interactive* variable, because its categories are formed by the 2 × 7 combinations of "sex" and "profession." The latter two variables also might have been treated as separate variables 13A (sex) and 13B (profession).

PRIMALS object scores depend upon the active variables only. A high object score on the first dimension means that the respondent believes in the effectiveness of social measures of prevention, does not believe in the effectiveness of penal measures, and has little fear. On this basis male respondents have an average score of .092, and females of −.084. This produces a discrimination measure of .008—clearly not significant. The discrimination measure is calculated as follows. There are 581 males and 635 females in the example. The discrimination measure then becomes

$$\{(.581)(.092)^2 + (635)(-.084)^2\}/1,216 = .008.$$

Variable 13B (profession) can be handled in the same way. The result is shown in Figure 1.1A. Because sex and profession are treated as two separate variables, the figure shows two parallel curves.

Figure 1.1B, on the other hand, gives the results when variable 13 is treated as interactive, as if it were one variable with 14 categories. Now the two curves are no longer parallel, and this indicates that sex differences are not the same for all professions. In the category "small business," males are more punitive and anxious than females, whereas in the categories "skilled labor" and "no profession" it is the other way round. One may conclude that the effects of sex and profession are not additive.

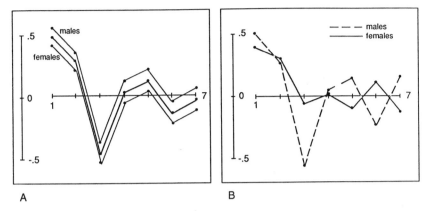

Figure 1.1. Category quantifications for variable 13 of Table 1.1. (A) Sex and age are treated as two separate variables, which results in parallel curves for males and females. (B) Sex and age are treated as one interactive variable.

One may attempt to explain the results in Figure 1.1B at face value. For instance, one may argue that for males, "no profession" has a different sociological meaning than it does for females. But one also may suspect that sex differences within a given profession depend on other factors. In the category "skilled labor," for instance, it could be true that females are on the average much younger than males. One also could pay attention to category frequencies. For instance, "no profession" is mentioned by 160 men, compared with 496 women, and "small business" is mentioned by 41 men, compared with 9 women.

Two conclusions can be drawn. The first is that passive treatment of a variable still makes it possible to find out whether there is a relation between object scores (based on active variables only) and a passive variable (which does not contribute to object scores). The second conclusion is that interactive treatment of two (or more than two) variables may reveal that the effects of those variables are not additive.

1.3.4 PCA Results

The second block of PRIMALS produces PCA results, before and after optimal quantification. First, the correlation matrices before and after can be displayed. For the example they are shown in Table 1.4 (after) and Table 1.5 (before). In these matrices the separate variables 13A (sex) and 13B (profession) have been included. As a result, Tables 1.4 and 1.5 each

TABLE 1.4 Matrix of Correlations After Transformation

	2	3	4	5	6	7	8	9	10	13A	13B
1	.12	.23	.40	.06	.32	.01	.07	.12	.06	.04	.08
2		.50	.19	.15	.18	.07	.06	.07	.00	.01	.06
3			.28	.31	.22	.11	.12	.11	.01	.08	.15
4				.11	.45	.09	.11	.15	.06	−.03	.13
5					.08	.08	.08	.11	.01	.03	.10
6						.06	.07	.12	.05	−.01	.07
7							.31	.13	.07	.12	.03
8								.19	.20	.26	.08
9									.22	.07	.06
10										−.04	−.06
13A											.22

contain 12 variables, so that there are also 12 solutions for PCA eigenvalues. The first two eigenvalues, with corresponding component loadings, are shown in Table 1.6. If passive variables 13A and 13B were not included, the first eigenvalue after transformation (equal to .2482) would have been the same as the first PRIMALS eigenvalue (equal to .2418). But adding the two passive variables 13A and 13B makes the first eigenvalue of the correlation matrix somewhat larger than the first PRIMALS eigenvalue (it cannot become smaller). Also, component loadings in PCA after transformation are not exactly the same as those given by PRIMALS.

In this example the differences between PCA results before and after transformation are quite small. This could be expected from the fact that the optimal transformations are almost linear. (There is a trivial difference with respect to the sign of component loadings of 13A and 13B before and

TABLE 1.5 Matrix of Correlations Before Transformation

	2	3	4	5	6	7	8	9	10	13A	13B
1	.10	.22	.41	.05	.32	.01	.07	.13	.07	−.04	−.02
2		.49	.17	.15	.17	.06	.07	.06	.00	−.02	−.01
3			.27	.29	.22	.11	.12	.11	.01	−.08	−.05
4				.10	.44	.07	.09	.14	.06	.04	−.04
5					.06	.09	.08	.11	.02	−.04	−.05
6						.05	.06	.12	.04	.02	.00
7							.30	.13	.07	−.12	−.07
8								.20	.20	−.26	−.15
9									.22	−.07	−.08
10										.04	.04
13A											.47

TABLE 1.6 PCA Component Loadings After and Before Transformation

	After		Before	
1	.546	−.268	.543	−.279
2	.525	−.216	.504	−.165
3	.674	−.164	.658	−.132
4	.655	−.287	.632	−.342
5	.387	.015	.371	.036
6	.583	−.292	.564	−.341
7	.304	.508	.315	.367
8	.380	.658	.410	.516
9	.382	.323	.399	.182
10	.179	.328	.201	.099
13A	.193	.516	−.232	−.703
13B	.290	.125	−.218	.614
Eigenvalue	2.482	1.498	2.405	1.672

after transformation. Apparently, the a priori quantification of these two variables is in the "wrong" order.)

Figure 1.2 gives a plot of the variables in terms of their component loadings, after transformation, on the first two PCA dimensions. The following tentative conclusions can be drawn:

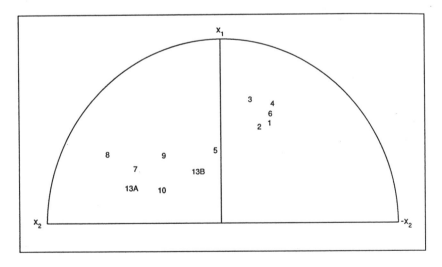

Figure 1.2. Component loadings for crime and fear example, from Table 1.6, with the direction of x_2 mirrored.

1. On the first dimension, object scores are correlated with "not anxious," with "no belief in the effectiveness of punitive measures," and with "belief in the effectiveness of social measures." There is also a small correlation with "profession" (in the sense that high object score on the first dimension tends to go together with "higher" professional group).
2. The second dimension indicates a distinction between the "fear" variables (7 to 10) and the "prevention" variables (1 to 6). Also, within the set of prevention variables the "social" variables (1, 4, and 6) tend to form a cluster distinct from the cluster of "punitive" variables (2, 3, and 5).
3. From the correlation matrix itself it can be seen that fear and prevention variables have small correlations with each other. In Figure 1.2 the same is revealed by the fact that the fear variables tend to form a bundle that is orthogonal to the bundle of prevention variables.
4. In principle, it is possible to add the 1,216 object points to Figure 1.2. Note, however, that the bulk of those points would be located far outside the framework of the figure.

1.3.5 Location of SC Points

Chapter 3 of the companion volume on theory explains that it is also possible to locate the SC points in a graph such as Figure 1.2. I shall illustrate this here for category 3 of variable 7 in the example. But first, in order to facilitate the discussion, let us define the *component point* of a quantified variable as the point with coordinates equal to the component loadings and the *component line* as the line through this point and the origin.

In the example we have 10 variables. It follows that for a complete PCA solution based upon the first quantification, the object points must be located on a 10-dimensional regular lattice. Moreover, because there are 6 variables with 5 categories and 4 variables with 3 categories, the possible category combinations define $(5^6)(3^4)$ lattice points, more than a million. It makes no sense, therefore, to attempt to sketch how this lattice projects onto a two-dimensional plane, such as the plane shown in Figure 1.2.

All object points in category 3 of variable 7 will be located in a 9-dimensional hyperplane of the lattice. Moreover, this hyperplane will be orthogonal to the component line of variable 7, and this line intersects the hyperplane at the SC point of category 3.

Coordinates of this SC point are found by multiplying the component point with the standard scores of the category. In the example, category 3 has MC of .310 (Table 1.3). The variance of the MC values of variable 7 is equal to the discrimination measure of this variable, in the example equal to .085. The standard score of category 3 therefore will be equal to $(.310)/(.085)^{1/2} = 1.06$. On the first two dimensions variable 7 has component loadings (.304, .508). Multiplication by 1.06 results in (.323, .540), the coordinates of the SC point.

Suppose for a moment that the component point were located in the plane of drawing. It then would be true that the hyperplane of category 3 projects on the figure as a single straight line, orthogonal to the component line, and passing through the SC point. However, in the example the component point of variable 7 is not located in the plane of drawing. This can be seen from the fact that the first two component loadings do not have a sum of squares equal to 1. It follows that the hyperplane of category 3 is not orthogonal to the plane of drawing, so that object points in this hyperplane will not project on a straight line. But if the sum of the squared loadings is close to 1, the object points will have projections close to a straight line. Similarly, objects in another category of variable 7 would also be located in a narrow strip (orthogonal to the component line and passing through the SC point of the category).

In the example, however, the sum of the squared first two component loadings is far below 1. The component line of variable 7 deviates quite a lot from the plane of drawing. Objects in the same category will project on wide strips, with a lot of overlap.

1.4 Second PRIMALS Example:
Socioeconomic Variables

1.4.1 Introduction

Data for this second example were made available by the group Energy and Environment at the University of Leiden. They are taken from a survey among 1,076 respondents. The survey was focused primarily on beliefs and attitudes with respect to the use of various sources of energy (gas, coal, nuclear) and the consequences for the environment. From this survey I have selected nine variables that in the survey itself were meant only to provide background information about the respondents. The variables are as follows:

1. profession (student, employed, retired, unemployed, unfit for work, housewife)
2. level of profession (in seven more or less increasing degrees)

3. level of education (in five increasing steps)
4. income (four increasing steps)
5. well-being (five steps)
6. age
7. sex
8. political preference
9. religious affiliation

For the present example we will not concentrate on substantive PRI-MALS results, but only on some special technical aspects. They will be demonstrated on the basis of four different PRIMALS solutions:

1. PRIMALS based upon only the first five variables (in the PCA block of PRIMALS, missing data replaced by object scores)
2. PRIMALS based upon only the first five variables, but missing data replaced by zeros
3. PRIMALS based upon only the first five variables, but with the last four variables as passive
4. PRIMALS over all nine variables taken as active

As a preliminary, it should be pointed out that the number of missing data on the first five variables is quite divergent: 3, 442, 0, 229, and 0, respectively. Actually, only variables 2 and 4 have so many missing data that they can play a role.

1.4.2 Optimal Quantification

About the category quantifications of the first five variables, it suffices to report that they are almost linear transformations of the a priori quantification—in the same order as given above for variables 2 to 5, and in reversed order for variable 1.

In variable 6 quantification decreases from young to old. In variable 7 men have larger quantification than women. Variable 8 (political parties) shows high quantification for VVD (the "conservative" party) and for "small left" (rather surprisingly), whereas PvdA (the large "left" party) scores low. On variable 9, "no religion" has a high score and all other categories relatively low.

In Solutions 1, 2, and 3, the quantifications on the first dimension are identical. The reason is, of course, that these three analyses differ only in their PCA block, and not in the first PRIMALS block for optimal quantification. In Solution 4 the quantification is somewhat different, although the differences are small. Because age and sex are active variables in Solution 4, it follows that the quantification of the first five variables pays more attention to implicit age differences (such as hidden in the categories "student" versus "retired" in variable 1) and to hidden sex differences (as in the category "housewife" in variable 1).

1.4.3 Discrimination Measures and Component Loadings

Discrimination measures and component loadings for the first two PCA dimensions (after transformation) are given in Table 1.7. The results on the second dimension are listed only for the sake of completeness—in the following discussion they will be ignored.

We start with a comparison of Solutions 1 and 2. If there were no missing data, these two solutions would be identical. In particular, discrimination measures would be the square of the component loadings. In the example, however, there are missing data, especially in variables 2 and 4, and so we find that discrimination measures are not the same as the squared component loadings.

Table 1.7 also shows eigenvalues. In the PRIMALS solution the eigenvalue is equal to the average of the discrimination measures, and in a PCA solution equal to the average of the squared component loadings. Typically, in Solution 1 the PCA eigenvalue of .627 is larger than the PRIMALS eigenvalue of .569. The reason is that in the PRIMALS solution missing data are treated as passive, whereas in PCA Solution 1 they are replaced by object scores, which gives a flattering image. In Solution 2, however, missing data are replaced by zeros. This has a lowering effect upon the component loadings, especially for Variables 2 and 4, where there are many missing data. As a consequence, the PCA eigenvalue of Solution 2 becomes smaller than the PRIMALS eigenvalue.

In Solution 3, the last four variables are treated in the PCA block as if they were active. As a consequence, component loadings of the first five variables becomes smaller than in Solution 1. In Solution 1 their sum of squares was *maximized* to the value of .627. In Solution 3 it reaches a value of only .601.

In Solution 4, the PRIMALS solution is based on all nine variables. This has the effect that for the first five variables the average discrimination measure becomes smaller than in Solution 3: It drops from .569 in Solution 3 to .489 in Solution 4. The same effect can be seen in the PCA solution, where the average square of the component loadings of the first five variables goes down from .601 in Solution 3 to .585 in Solution 4. On the

TABLE 1.7 Socioeconomic Variables: Discrimination Measures and Component Loadings in Four Analyses

Variables	Discrimination Measure	Component Loadings	
Solution 1			
1	.363	.584	.785
2	.500	.915	.015
3	.833	.846	−.027
4	.419	.790	−.128
5	.728	.786	−.443
Eigenvalue	.569	.627	.166
Solution 2			
1	.363	.503	.842
2	.500	.766	−.267
3	.833	.839	.106
4	.419	.646	−.113
5	.728	.798	−.295
Eigenvalue	.569	.520	.178
Solution 3			
1	.363	.642	.600
2	.500	.894	−.079
3	.833	.825	−.124
4	.419	.766	−.220
5	.728	.725	−.459
Eigenvalue	.569	.601	.128
6	.115	.413	.379
7	.106	.412	.619
8	.111	.598	−.263
9	.012	.150	.241
Solution 4			
1	.539	.711	.531
2	.342	.903	.026
3	.747	.804	−.259
4	.343	.749	−.182
5	.476	.629	−.571
Eigenvalue	.489	.585	.142
6	.324	.541	.219
7	.206	.452	.627
8	.210	.618	−.208
9	.058	.218	.137

other hand, the last four variables have larger component loadings in Solution 4 than in Solution 3. Their average square increases from .180 in Solution 3 to .231 in Solution 4. A similar effect is found for the discrimination measures of these four variables in the PRIMALS solution. Their average rises from .086 in Solution 3 to .200 in Solution 4.

1.5 Conclusions:
When PRIMALS, and What Are the Alternatives?

The first conclusion is that PRIMALS can be useful if there is reason to assume that all categorical variables measure "the same thing" and that they are, at least to some extent, dependent on one underlying common variable. The PRIMALS quantification then assures that the variables are quantified in such a way that they become as similar to each other as possible.

PRIMALS therefore cannot be recommended for a set of variables that may be assumed to measure different things (even if it may be true that the variables are somewhat correlated). In this situation it may be advisable to try HOMALS (which gives different quantifications in subsequent dimensions) or to try one of the varieties of PRINCALS (for the same reason). Another option is to restrict PRIMALS to a subset of active variables, with other variables as passive (as in the example given in Section 1.4). The active set should contain variables that probably have a lot in common, whereas variables that seem to measure different things are treated as passive.

If there are clearly two different subsets of variables, CANALS might be recommended. Section 1.3 gives an example, with an a priori distinction between "prevention" variables and "fear" variables. Or, if one can make a priori distinctions among more than two sets of variables, application of OVERALS should be considered.

2 HOMALS

2.1 Introduction

The acronym HOMALS is based on the abbreviations HOM, for analysis of *hom*ogeneity, and ALS, for *a*lternating *l*east *s*quares. The term will be used for the specific technique of multiple optimal quantification, as well as for the corresponding computer program.

The method has also been given many other names, the best known of which is *multiple correspondence analysis.* This is the name used by Benzécri et al. (1973), whose publications have become very popular in France as well as in English-speaking countries (see Greenacre, 1984). Other names are *dual scaling, method of reciprocal averaging, linearization of regression,* and *seriation.* One reason there are many different names is that the method has been "discovered" or "invented" inde-

18

TABLE 2.1 Illustration of Concept of Homogeneity

X			Y			Z		
10	6	5	7	7	7	3	−1	−2
4	2	6	4	4	4	0	−2	2
−3	2	−2	−1	−1	−1	−2	3	−1
−4	−3	−5	−4	−4	−4	0	1	−1
−7	−7	−4	−6	−6	−6	−1	−1	2

NOTE: In X the observed scores are shown as deviations from their column means. In Y they are replaced by their row means. Z shows the difference between X and Y.

pendently by many people. These researchers often focused on one special property of the method without being aware of other properties. Another reason is that the method was sometimes invented specifically for application within one particular field of study, without awareness that virtually the same method already was known in different fields.

The concept of homogeneity will be clarified with the help of an example. Table 2.1 gives a 5 × 3 matrix of observations on $n = 5$ objects and $m = 3$ variables. At the left of this table we have a matrix X in which the three variables have zero column means. Replace elements in the rows of X by their row means. This gives matrix Y. Define Z as the difference matrix $Z = X - Y$. Rows and columns of Z have zero means. This property is meant when Z is said to be "double centered."

In Table 2.1 the matrix Y may be called *perfectly homogeneous* because its rows differ only in their means. Further, if rows of Y were replaced by row means, nothing would change. On the other hand, matrix Z is *nonhomogeneous*. If we replace elements of Z by their row means, we will obtain a zero matrix.

Borrowing terms from analysis of variance, we could say that the sum of the squared elements in X gives the *total* sum of squares T. The sum of squares of the elements in Y depends only on differences between rows, and therefore is called the *between* sum of squares B. The sum of squares of the elements in Z depends only on differences within rows, and is called the *within* sum of squares W. Matrix X would be perfectly homogeneous if Z were a zero matrix, and therefore if $B = T$.

In the example, matrix X is not perfectly homogeneous. A measure of the amount of homogeneity is the ratio B/T. This ratio is equal to 1 if X is perfectly homogeneous, and equal to 0 if rows of X have identical zero means.

Let us now go back to optimal quantification. Suppose that there is no a priori quantification, and that matrix X does not contain numerical values

TABLE 2.2 Example Used in the Text to Show That Optimal Quantification
Leads to Maximal Homogeneity

	X			y
a p v	.748	−.622	.264	.130
a q u	.748	.656	.960	.788
b r v	.877	−.407	.264	.247
b r v	.877	−.407	.264	.247
b q u	.877	.656	.960	.831
b q u	.877	.656	.960	.831
c p w	−.834	−.622	−1.224	−.893
c r u	−.834	−.407	.960	−.094
c r w	−.834	−.407	−1.224	−.822
c r w	−.834	−.407	−1.223	−.822
c q w	−.834	.656	−1.224	−.467
c q v	−.834	.656	.264	.029

but only labels for the different categories of each variable. Table 2.2 gives
an example, with the same data as in Table T.3.8. The matrix at the left of
Table 2.2 gives category labels only. The matrix in the middle changes
these labels into their optimal quantification (in agreement with Table
T.4.5). This matrix in the middle may be called X. At the right of Table 2.2
we find a column y with the averages of the rows of X. These averages can
be interpreted as object scores. They are proportional to the object scores
given in Table T.4.5, with proportionality factor equal to $\lambda^2/m = 1.84/3 =$
.613.

In Table 2.2 the elements of X have sum of squares equal to $T = 22.08$.
If elements of X are replaced by their row means, given in column y, the
sum of squares becomes equal to $B = 13.54$. The essential point is that
categories have been quantified in such a way that the ratio B/T is maximized, and has value $13.54/22.08 = .613 = \lambda^2/m$. In other words, the effect
of optimal quantification is that the homogeneity of X is maximized. It
will be obvious that this solution also maximizes the ratio B/W, and that
it minimizes the ratio W/T.

The second solution, on the next dimension for optimal multiple quantification, assures that the B/T ratio is maximized *under the condition* that
object scores must be produced that are uncorrelated with the object scores
on the first dimension.

The example illustrates that the approach by which homogeneity is
maximized, in terms of B/T ratios, comes to the same thing as optimal
multiple quantification, the theory of which is explained in Chapter 4 of
the companion volume on theory, especially in Section T.4.6. That section

also gives algebraic and geometric details of the HOMALS solution, and there is no need to repeat those details here, except for the following remarks.

First, HOMALS requires that successive solutions for object scores must be uncorrelated among each other. But this does not imply that successive quantifications of the same variable will be uncorrelated.

Second, there is an exception to this rule. If HOMALS is applied to a situation with only two categorical variables, successive quantifications of these two variables will be uncorrelated. In fact, the program ANACOR has been developed especially for the situation with only two categorical variables, and gives uncorrelated quantifications of them.

Third, a binary variable (which has only two categories) can be quantified in only one way. Successive quantifications of such a variable therefore are perfectly correlated (unless such a variable obtains zero quantification on one or more dimensions).

The fourth remark is more general. If a variable has k_j categories, the category points will be restricted to a space with $(k_j - 1)$ dimensions. For instance, the $k_j = 2$ categories of a binary variable will always be on a straight line, and therefore are confined to a one-dimensional space. If $k_j = 3$, the three categories always will be located in a flat plane, and are restricted to a two-dimensional space. It follows that a variable with k_j categories never can have more than $(k_j - 1)$ uncorrelated quantifications. In fact, if there are more than $(k_j - 1)$ HOMALS solutions, there must be linear dependence among the quantifications of the variable.

2.2 When Is HOMALS Useful?

First of all, when all variables are binary, results of HOMALS will be the same as those obtained by classical PCA, no matter what a priori quantification was chosen for the variables.

HOMALS is useful when it produces interesting graphs of results. Let us look at such a two-dimensional graph based upon the first two HOMALS quantifications.

1. The graph contains n object points, with object scores as their coordinates.
2. The graph contains for each category of the m variables an MC point. These points are defined as the center of gravity of objects within the category.
3. The graph has two major characteristics: (a) In a good solution, object points will be close to their MC points (more precisely, the sum of squares of distances between object points and their corresponding MC points is minimized), and

(b) category points will have large spread, because the solution maximizes the sum of squared distances between MC points and the origin.

The two properties of the graph noted in the third point above are closely linked for the following reason. On each dimension the sum of squared distances of object points to the origin is equal to n (given that the object points have zero mean and unit variance—which may not be exactly true if there are missing data that are treated as passive). For the two dimensions combined, the sum of squared distances to the origin therefore will be equal to $2n$. The two sums of squares mentioned above are additive to the total sum of squares of $2n$. It follows that minimization of the first sum of squares (which refers to distances between object points and their MC points) implies maximization of the second sum of squares (distances of MC points to the origin).

In the graph, it will be interesting to see for which variables the MC points have large spread. Categories discriminate between objects better to the extent that they are farther away from each other. This spread is indicated in the discrimination measure. A small discrimination measure implies that on that dimension the category points of a variable are close to the origin, and therefore that distances between object points and their category points will be relatively large. On the other hand, if a discrimination measure is large, category points are far away from each other on that dimension, and object points are close to their category points. In this way the graph shows for each dimension which variables are effective and which are not.

The overall success of a HOMALS solution is expressed in the eigenvalues. They give, for each dimension, the average value of the discrimination measures.

2.3 Some Special Remarks

2.3.1 More Than Two Dimensions

A HOMALS solution may allow for more than two dimensions. In fact, if there are no passive missing data, the number of dimensions has upper bound ($\sum k_j - m$), where m is the number of variables, and k_j the number of categories in the jth variable. This issue is discussed more fully in Section T.6.7.

HOMALS solutions are *nested*. This means that a HOMALS solution in p dimensions is the same as that for the first p dimensions in a HOMALS solution with more than p dimensions. In other words, increasing the number of dimensions does not require a revision of the quantification in preceding dimensions.

2.3.2 HOMALS Not Always Fair

Suppose that on the first HOMALS dimension we obtain a relatively large eigenvalue λ_1^2. This does not guarantee that all individual discrimination measures must be large. It may happen that the solution is dominated by the large discrimination measures of only some variables, whereas other variables are neglected in the sense that their discrimination measures are close to zero.

Such neglected variables will become more dominant on subsequent dimensions. The reason is indicated in Section T.6.3, where it is argued that a variable with k_j categories will have sum of discrimination measures equal to $(k_j - 1)$ if all possible dimensions for multiple quantification are taken. It follows that a variable with near-zero discrimination measures on the first dimensions must on the remaining dimensions have a sum of discrimination measures close to $(k_j - 1)$.

Another point is that it seems to be a very nice result if a large HOMALS eigenvalue is found, close to 1. Such a result implies that the different variables are closely related to each other, in the sense that they sort out objects in the same way. However, this becomes a very trivial result if different variables contain categories that overlap in terms of how they are defined. For instance, one does not need to analyze data (or even to collect them) to find out that "grandfathers" are "male," that they "have children," and that they tend to belong to "older age" groups. In fact, there is a kind of paradox here, in that a researcher will feel disappointed when there are no large eigenvalues, but a very large eigenvalue may turn out to be equally disappointing.

2.3.3 Leading Category Quantifications

In the companion volume on theory, a distinction is made between quantification of categories in terms of MC points (multiple coordinates points) or in terms of LC points (leading category points). The MC point of a category is defined as the center of gravity of the scores of objects within the category. However, there are situations in which one might prefer a different kind of graph, a graph in which an object point appears as the center of gravity of the categories that apply to the object. MC points then must be replaced by different category points, which are here referred to as LC points.

As an example, suppose we have collected data about the countries in Latin America based upon categorical variables related to the economy of those countries. There are not many Latin American countries, and so the number of objects is relatively small. The number of variables might be much larger. In a graph of the results, it might be much more revealing to plot countries as centers of gravity of categories applying to them. In such a graph, Country X appears as "representative" of a country with specific economic properties, perhaps close to Country Y, which has similar properties, but far away from Country Z, which has quite different economic characteristics.

HOMALS is an application of optimal multiple quantification and therefore obeys the principle of reciprocal averaging. The beauty of this principle is that it implies that for a given HOMALS dimension coordinates of MC points are essentially the same as those of LC points. They differ only in terms of how they are normalized. This means in fact that LC values are proportional to MC values. In the companion volume on theory it is illustrated, in various places, that the proportionality factor is equal to the HOMALS eigenvalue of the dimension. The simple relation is that coordinates of LC points are obtained if the coordinates of MC points are divided by this eigenvalue.

(To avoid confusion, it must be remarked that in the volume on theory an eigenvalue λ^2 is defined as the *sum* of the discrimination measures, and that the proportionality factor between LC and MC is given as λ^2/m. However, the HOMALS program defines an eigenvector as the *average* of the discrimination measures, and therefore equal to λ^2/m.)

2.4 Structure of the HOMALS Program

The HOMALS program gives optimal quantification of the categories of each variable in p dimensions. The number p is prescribed by the user. Usually, the user keeps p quite small: $p = 2$ or $p = 3$, for instance. The program gives for each dimension the discrimination measures of the variables and their average (the HOMALS eigenvalue). The program gives the quantifications of the categories (standardized or in terms of MC values) on each dimension, and can print, on request, the object scores. Furthermore, the program can print, on request, various graphs. Illustrations of such graphs will be provided later, in the discussion of numerical examples.

The HOMALS program does not (as PRIMALS does) produce results of PCA applied to the quantified variables on each dimension. In fact, one should realize that each dimension of HOMALS gives a different quantification of the variables. As a result, one can obtain a complete m-dimensional PCA solution for each of the HOMALS quantifications. This results in an enormous proliferation of PCA solutions.

To illustrate, suppose there are $m = 3$ categorical variables, each with 3 categories. Assume that there are no missing data. Then there are $2 + 2 + 2 = 6$ possible HOMALS solutions. Each of them gives a 3×3 matrix of correlations between the quantified variables. Each of these six correlation matrices has three PCA solutions, so that we end up with $6 \times 3 = 18$ PCA dimensions.

Let the subsequent correlation matrices be numbered, from R_1 to R_6. The first HOMALS solution will be identical to the first PCA solution of R_1. The second HOMALS solution will be identical to one of the PCA solutions of R_2, but it need not be the first of them—it could also be the second. We see a sort of shift. As the number of the HOMALS solution increases from 1 to 6, the number of the corresponding PCA solution goes up from 1 to 3. The last HOMALS solution will correspond to the last PCA solution of the last correlation matrix.

2.5 First HOMALS Example:
Dental Patterns of Mammals

This first example, taken from Van de Geer (1985), concerns the dental patterns of 66 different mammals. For each animal, the number of teeth is counted on the following eight variables:

1. top incisors
2. bottom incisors
3. top canines
4. bottom canines
5. top premolars
6. bottom premolars
7. top molars
8. bottom molars

The data are given in Table 2.3. In the HOMALS analysis the numbers in the cells of this matrix are treated merely as labels—the variables are treated as nominal. The table also classifies the mammals in terms of their biological families and orders. The main question is whether HOMALS will recover the classification in orders.

TABLE 2.3 Dental Patterns of Mammals

Objects	1	2	3	4	5	6	7	8	Family	Order
					Variables					
Opossum	5	4	1	1	3	3	4	4	Didelphidae	B
Hairy tailed mole	3	3	1	1	4	4	3	3	Talpidae	I
Common mole	3	2	1	0	3	3	3	3		
Star-nosed mole	3	3	1	1	4	4	3	3		
Brown bat	2	3	1	1	3	3	3	3	Microchiroptera	V
Silver haired bat	2	3	1	1	2	3	3	3		
Pygmy bat	2	3	1	1	2	2	3	3		
House bat	2	3	1	1	1	2	3	3		
Red bat	1	3	1	1	2	2	3	3		
Hoary bat	1	3	1	1	2	2	3	3		
Lump nosed bat	2	3	1	1	2	3	3	3		
Armadillo	0	0	0	0	0	0	8	8	Dasypodidae	T
Pika	2	1	0	0	2	2	3	3	Ochotonidae	H
Snowshoe rabbit	2	1	0	0	3	2	3	3	Leporidae	
Beaver	1	1	0	0	2	1	3	3	Castoridae	K
Marmot	1	1	0	0	2	1	3	3	Sciuridae	
Groundhog	1	1	0	0	2	1	3	3		
Prairie dog	1	1	0	0	2	1	3	3		
Ground squirrel	1	1	0	0	2	1	3	3		
Chipmunk	1	1	0	0	2	1	3	3		
Gray squirrel	1	1	0	0	1	1	3	3		
Fox squirrel	1	1	0	0	1	1	3	3		
Pocket gopher	1	1	0	0	1	1	3	3	Geomydae	
Kangaroo rat	1	1	0	0	1	1	3	3	Heteromyidae	
Pack rat	1	1	0	0	0	0	3	3	Cricetidae	
Field mouse	1	1	0	0	0	0	3	3		
Muskrat	1	1	0	0	0	0	3	3		
Black rat	1	1	0	0	0	0	3	3		
House mouse	1	1	0	0	0	0	3	3		
Porcupine	1	1	0	0	1	1	3	3	Erethizontidae	
Guinea pig	1	1	0	0	1	1	3	3		
Coyote	1	3	1	1	4	4	3	3	Canidae	R
Wolf	3	3	1	1	4	4	2	3		
Fox	3	3	1	1	4	4	2	3		
Bear	3	3	1	1	4	4	2	3	Ursidae	
Civet cat	3	3	1	1	4	4	2	2	Viverridae	
Raccoon	3	3	1	1	4	4	3	2	Procyonidae	
Marten	3	3	1	1	4	4	1	2	Mustelidae	
Fisher	3	3	1	1	4	4	1	2		

continued

TABLE 2.3 Continued

Objects	1	2	3	4	5	6	7	8	Family	Order
				Variables						
Weasel	3	3	1	1	3	3	1	2		
Mink	3	3	1	1	3	3	1	2		
Ferret	3	3	1	1	3	3	1	2		
Wolverine	3	3	1	1	4	4	1	2		
Badger	3	3	1	1	3	3	1	2		
Skunk	3	3	1	1	3	3	1	2		
River otter	3	3	1	1	4	3	1	2		
Sea otter	3	2	1	1	3	3	1	2		
Jaguar	3	3	1	1	3	2	1	1	Felidae	
Ocelot	3	3	1	1	3	2	1	1		
Cougar	3	3	1	1	3	2	1	1		
Lynx	3	3	1	1	3	2	1	1		
Fur seal	3	2	1	1	4	4	1	1	Otariidae	
Sea lion	3	2	1	1	4	4	1	1		
Walrus	1	0	1	1	3	3	0	0	Odopenidae	
Gray seal	3	2	1	1	3	3	2	2	Phocidae	
Elephant seal	2	1	1	1	4	4	1	1		
Peccary	2	3	1	1	3	3	3	3	Tayassuidae	E
Elk	0	4	1	0	3	3	3	3	Cervidae	
Deer	0	4	0	0	3	3	3	3		
Moose	0	4	0	0	3	3	3	3		
Reindeer	0	4	1	0	3	3	3	3		
Antelope	0	4	0	0	3	3	3	3	Bovidae	
Bison	0	4	0	0	3	3	3	3		
Mountain goat	0	4	0	0	3	3	3	3		
Musk-ox	0	4	0	0	3	3	3	3		
Mountain sheep	0	4	0	0	3	3	3	3		

NOTE: Numbers indicate how many teeth the animal has for each of the eight types described in the text. The far right column identifies order: B = Marsupiala, I = Insectivora, V = Chiroptera, T = Edentata, H = Lagomorpha, K = Rodentia, R = Carnivora, E = Artiodactyla.

Table 2.3 shows that there are three animals with unique dental patterns. One of them is the opossum, unique in category 5 of variable 1 and in category 4 of variables 7 and 8. Another unique animal is the armadillo, unique in category 9 of variables 7 and 8, and in category 0 of all other variables. The third one is the walrus, unique in category 1 of variables 7 and 8.

HOMALS is sensitive to such unique objects, called *outliers*. There will be dimensions on which a unique category obtains extreme quantification, whereas all other categories are merged. In fact, the first HOMALS dimension shows only that the opossum is an outlier. When the opossum

TABLE 2.4 Dental Patterns of Mammals: HOMALS Discrimination Measures

Variables	Discrimination Measures	
	Dimension 1	Dimension 2
1	.864	.866
2	.819	.856
3	.813	.003
4	.821	.047
5	.657	.656
6	.726	.718
7	.664	.027
8	.584	.015
Eigenvalue	.744	.399

is dropped from the analysis, a first HOMALS dimension appears on which the armadillo is the outlier. Dropping this object, too, results in a first dimension with the walrus as an outlier.

In the following example, therefore, these three outliers are discarded, and HOMALS is applied to the remaining 63 objects. Table 2.4 shows the discrimination measures for the first two dimensions, which have eigenvalues of .744 and .399, respectively.

Objects are plotted in the graph of Figure 2.1. In this graph the mammals are labeled with the letters giving their orders (shown in the rightmost column of Table 2.3 and explained in Table Note). In each case, the number next to the letter gives the frequency of objects with the same dental profile. Close to this number the dental profile itself is given. For instance, in the right lower corner of the graph we find an object with label K5, indicating that there are five rodents with identical profile (11000033).

Figure 2.1 is dominated by a sort of triangle, formed by the orders E, K, and R. Carnivora R (at the left) have many teeth of all sorts except molars. Artiodactyla E and rodents K (at the right) have many molars but no canines. But E tend to have more bottom incisors, whereas K have some top incisors.

It is clearly seen in Figure 2.1 that the six orders tend to form clusters. In a way this seems to be quite natural, because the HOMALS criterion requires that object points should be close to their MC points. Nevertheless, HOMALS is not the same as cluster analysis. The latter requires a quantification of objects in such a way that they fall apart into separate "clouds." The criterion is that pairs of objects within the same cluster have smaller interdistances than pairs of objects from different clusters. This is not exactly the same as the HOMALS criterion. But if a HOMALS solution

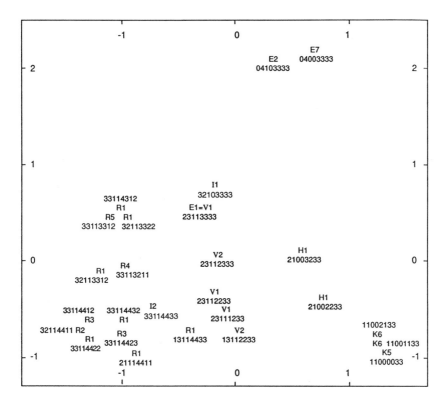

Figure 2.1. HOMALS solution for dental patterns of mammals, Table 2.3. The 26 different dental profiles are located at the position where a capital identifies the order of the mammals. To the right of the capital is the number of mammals with identical profile; nearby is the profile itself.

turns up with definite clusters, this result will probably be quite similar to that of a genuine cluster analysis.

2.6 Second HOMALS Example: Parliamentarians

In this second example the objects are the 75 members of the First Chamber of the Dutch Parliament and the 150 members of the Second Chamber, in September 1982. The data are reported in CBS (1983). (See the Appendix for more information about the Dutch parliamentary system and Dutch political parties.)

TABLE 2.5 Parliamentarians: Solution of First Two HOMALS Dimensions

Discrimination Measures Variables	*Dimension 1*	*Dimension 2*	*Sum*	*Maximum*
1 Chamber	.455	.066	.521	1
2 Age	.716	.542	1.258	5
3 Sex	.012	.510	.523	1
4 Party	.421	.352	.773	11

Category Quantifications	*Dimension 1*	*Dimension 2*	*Frequency*
First Chamber	−.95	.36	75
Second Chamber	.48	−.18	150
<30	2.54	1.60	3
30-34	.93	3.12	8
35-44	.67	−.24	73
45-54	.07	−.52	76
55-64	−.83	.45	57
≥65	−2.80	.17	8
Male	−.05	−.34	183
Female	.23	1.49	42
CDA	−.75	−.12	73
Centrum P	.82	−2.22	1
CPN	.94	2.76	4
D'66	.30	1.22	10
EVP	1.30	1.67	1
GPV	1.81	−1.63	3
PPR	1.18	.05	3
PSP	2.17	1.53	3
PvdA	.22	−.27	75
RPF	1.31	−1.93	2
SGP	−1.46	−.37	4
VVD	.44	.19	48

The example includes four nominal variables. The first variable is binary, and sorts objects according to whether they are members of the First or the Second Chamber. The second variable categorizes objects with respect to age, the third with respect to sex, and the fourth with respect to political party. Table 2.5 gives details about marginal frequencies, and also about HOMALS results in two dimensions. There are no missing data.

HOMALS results for this example are graphed in four figures. Figure 2.2 shows object points labeled with 1 or 2, depending upon whether they belong to the First or Second Chamber. One should keep in mind that a

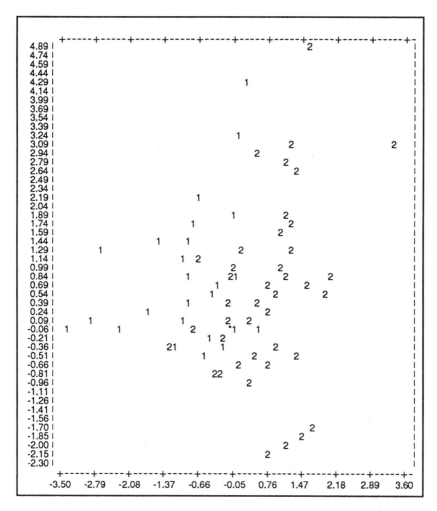

Figure 2.2. Parliamentarians. Object scores in HOMALS solution, with label 1 for members of First Chamber, label 2 for Second Chamber.

point may stand for more than one parliamentarian. The figure shows that the two categories are quite well distinguished. Members of the First Chamber are found primarily at the left of the figure, and those of the Second Chamber at the right. Nevertheless, the discrimination measure for this variable on the first HOMALS dimension, equal to .455, is not very large. This illustrates that a good discrimination between the categories of

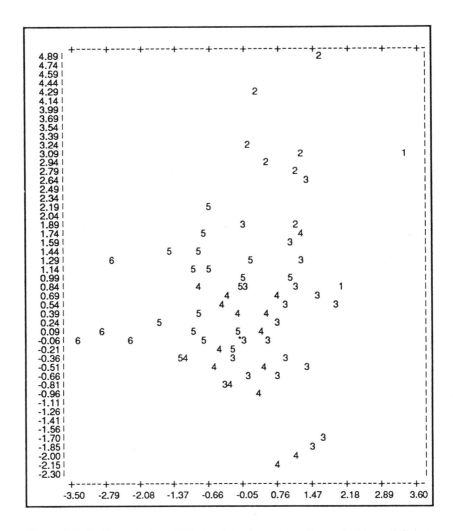

Figure 2.3. Parliamentarians. Object points at same location as in Figure 2.2, but now with labels 1 = youngest to 6 = oldest age group.

a variable does not necessarily imply that the discrimination measure must be very high.

Figure 2.3 plots objects at exactly the same locations as in Figure 2.2. The difference is in the labels: They now refer to the age categories, from 1 = relatively young to 6 = relatively old. Again we find that objects are

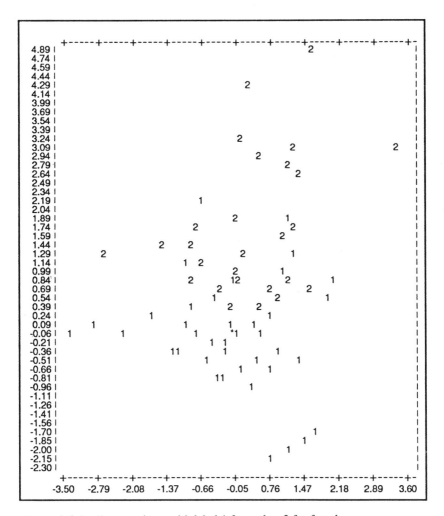

Figure 2.4. Parliamentarians, with label 1 for males, 2 for females.

in fairly distinct regions, with young parliamentarians (1 and 2) at the upper right and older ones (5 and 6) at the left of the figure.

Figure 2.4 has labels according to sex. The general pattern is that women (labeled 2) are in the upper half of the figure, and men (labeled 1) are in the lower.

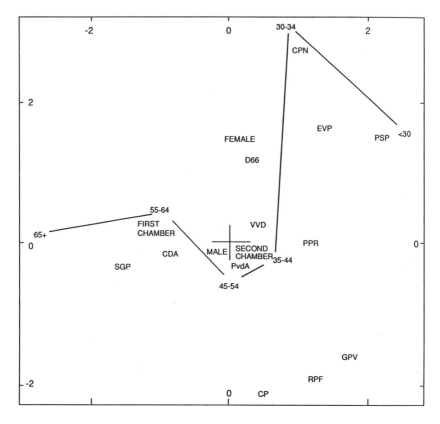

Figure 2.5. Parliamentarians. MC points.

Figure 2.5, finally, shows the MC points of the categories of all four variables. As might be expected, small political parties obtain more extreme quantification than do larger ones. Inspection of all these figures together shows that the smaller parties at the political left (CPN, EVP, PSP) are represented by parliamentarians who are relatively young, relatively more often female, and especially in the Second Chamber. Representatives of the smaller parties at the political right (GPV, RFP, CP) more often are males of middle age. SGP and CDA are relatively "overrepresented" in the First Chamber; their members tend to be older.

But there also are exceptions. In the upper right of the figures we find, for example, a young female CPN representative in the First Chamber. And

TABLE 2.6 Discrimination Measures of Background Variables in Crime and Fear Example

Variables	Dimension 1	Dimension 2
11 Religion	.304	.523
12 Voting	.394	.557
13A Sex	.373	.135
13B Profession	.486	.248
14 Age	.123	.203
Eigenvalue	.336	.333

in the middle of the plots we find a stray object in age category 5: a male representative of PSP in the Second Chamber.

The data for this example refer to the situation in September 1982. It would be interesting to compare them with results obtained in other years.

2.7 Third HOMALS Example: Background Variables

This third example makes use of the background variables mentioned in Table 1.1 as being part of the survey about crime and fear. For the present example variable 13 is split into two separate variables 13A (sex) and 13B (profession). The reason the example is given is that many researchers are inclined to apply HOMALS to such background variables, in spite of the fact that there will be little coherence among them. The HOMALS analysis then becomes a sort of shotgun approach. I would like to illustrate what may happen in such a situation.

Table 2.6 gives discrimination measures and eigenvalues for two dimensions. The discrimination measures are rather low, which indicates that the background variables certainly do not measure "the same thing." Figure 2.6 is a plot of the MC points. There is no convincing structure in them. Nevertheless, there are a few starting points for an attempt at interpretation.

First, category 5 (no religion) of variable 11 seems to be quite distinct from the other categories of this variable. Objects in category 5 tend to vote for PSP or to abstain from voting, and they tend to be male, relatively young, skilled workers or lower-rank employees. This is contrasted with objects who mention some religion, vote for some denominational party, are more often female, and have no profession.

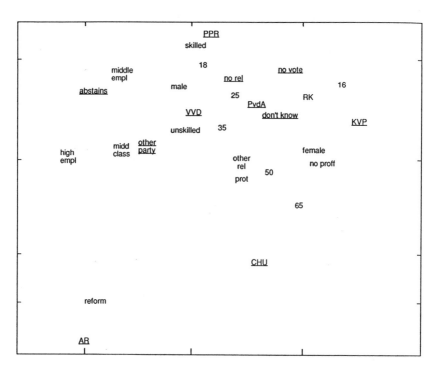

Figure 2.6. Crime and fear. Plot of the quantified categories of the background variables.

Second, we also find that objects who are Roman Catholic tend to vote KVP and tend to be younger, more often female, and without profession. They can be contrasted with reformed Protestant objects, who vote AR, are not so young, and belong to "higher" professions.

Nevertheless, it is inevitable that we still have some questions. What precisely is the sociological situation with respect to those young Roman Catholic females without profession? Is it not trivial that Roman Catholics more often give their votes to the Roman Catholic KVP, whereas Protestants vote for the Protestant AR? Is the result that the category "no religion" seems to be important not caused mainly by the tendency of HOMALS to exaggerate the quantification of categories with small marginal frequency?

We may conclude that one should not expect to obtain a lot of insight from a HOMALS analysis of a loose set of background variables.

2.8 Fourth HOMALS Example: Seriation

Data in this example refer to 59 graves dating from the Roman period and found in a graveyard near Münsingen-Rain in Switzerland. Such graves often contain various objects, such as remnants of earthenware decorated in different styles. In archaeology such objects are called *artifacts*.

The problem is whether it is possible to rank the graves on the basis of similarities between artifacts found in different graves. Obviously, one may expect that such a ranking will agree with an ordering of the graves with respect to their age. In archaeology this is known as the *seriation problem*.

Table 2.7 gives the Münsingen-Rain data, with the graves (59 rows) and the artifacts (columns) ordered on the basis of the HOMALS solution. The graves are labeled from A (oldest) to U (most recent) (this ordering is taken from Hudson, 1968).

Psychologists will recognize the formal identity between the seriation problem and Guttman's scaling method called *parallelogram analysis*. A typical example would be that respondents are given a list of the names of six politicians and are asked to pick out those two politicians they might favor with their vote. Assume that the six politicians can be ordered on one dimension, running from "conservative" to "liberal," and this is also the dimension that underlies the preferences of the voters. The raw data could be collected in an $n \times 6$ matrix, with 6 columns for the politicians, and n rows for the respondents. Each row will contain two elements equal to 1 (in the columns of the two politicians picked out by the respondent) and four elements equal to 0 (in the other columns). If the assumption above were true, it should be possible to arrange rows and columns of the matrix in such a way that the elements marked 1 appear in the shape of a parallelogram. This means that none of the rows should have an element 0 between the two 1s, and that the position of the two 1s shifts gradually from the extreme left in the top rows of the matrix to the extreme right in the bottom rows. The basic idea behind this solution is the same as in the method of "unfolding" described by Coombs (1964).

The example illustrates that archaeologists could have solved their seriation problem earlier if they had been familiar with methods developed in psychology. Situations such as this happen quite often. Scientists are not always aware that they are coping with problems that are formally identical to problems already addressed in other disciplines.

But let us return to Table 2.7. A basic question is how to treat the zeros in the table. There is no problem about the cells with unit entry. It cannot be denied that graves are more similar to the extent that they contain the

TABLE 2.7 Münsingen-Rain Data, Permuted Toward Parallelogram Structure on the Basis of HOMALS Solution

```
U   111
U   1.1
U   1...1
U    1111.1
U     11.11
U     111..1
U       1.1
U     1.11..1
U       1.11........................1
O           1.1111111.1..11......1..1
P            1111111..............1
R       1...1.....1..1.............1..1
T        1.1.....1.......1...1.1.....1
O            1..1.111...........1
R       1....11...................1
O            11.111.1.111..1..1.1
O            1......11..1...1
L            1..111....1.1...1
N            1..111.1..1...1.1
N              11..1..1.1
J                1.1
I                1.11
S             1........1
J             111......1
K            1.1.111......1
K            1.111......1
H           1..1.1.11.111
J               1.1.........1
I             11.1..1.1.1..1
G               11
G             1.1.111111
H               1....1
G             1.1.1111
H           1..1.11111..1.1
H         1.............1
F                 1...1.1
I         1..........1.1...1
R           1....................1
E                1..1
E                11..11....1
E                1......1
E                1.1111.....11
E                  11.1111
E                  1.111.1
E                   1..1
```

TABLE 2.7 Continued

D	1...1......11
A	1...1
A	1...1
C	1.11111....1
A	1...11
B	1............1111111.1
A	11
A	11.....1
A	11....1..1
A	11.1..11.1
A	1.....1..1
A	11......11
A	1..11...1
A	1.1

NOTE: Graves (rows) are indicated by capitals that refer to an independent alphabetical ordering from A = oldest to U = most recent.

same artifacts. However, are graves more similar to the extent that they do not contain the same artifacts? This question becomes even more pertinent if one realizes that absence of artifacts does not necessarily imply that they were never present. They may have been stolen, or perhaps they may have disintegrated. Also, an artifact may be missing from two graves just because there is a great age difference between them. The conclusion is clear: There is no reason to assume that graves are similar to the extent that the same artifacts are missing.

It follows that one should also look at the results of a HOMALS analysis in which the zeros of Table 2.7 are treated as passive missing data. Figures 2.7 and 2.8 picture the HOMALS results. Figure 2.7 is based on an analysis in which the zeros are treated as an independent category. The results are not very illuminating. In contrast, Figure 2.8 shows results when zeros are treated as passive missing data. Now the result is quite clear. Graves are ordered almost perfectly from U to A, albeit this ordering follows the shape of a horseshoe. Where does this horseshoe come from, assuming that age of the graves is the only underlying dimension? This question will be picked up in Section 2.10.

The conclusion from the example is that if one of the categories of a variable stands for absence of any of the other categories, it may be advisable to deal with this category as if it contained missing data, to be treated as passive.

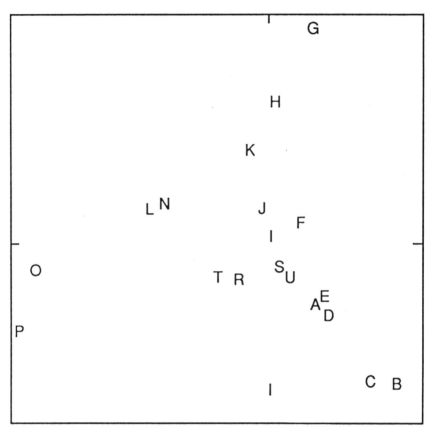

Figure 2.7. Münsingen-Rain data. HOMALS solution if absence of an artifact is treated as an independent category.

2.9 Reversed Indicator Matrix

What will be the effect if in an indicator matrix the roles of objects and variables is interchanged? As an example, take data about five persons who responded to four items, each with three categories. If we treat persons as objects and items as variables, the indicator matrix will have 5 rows and 4×3 columns. But the reversed indicator matrix will have 4 rows (for items treated as objects) and 5×3 columns (for five persons with three categories each).

Such a reversed indicator matrix will produce a good HOMALS solution if its columns are similar. This means that a subset of items allocated by one person to the same category also obtains an identical response from

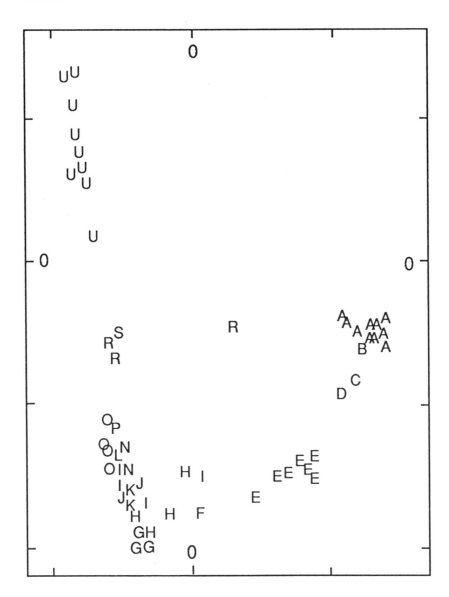

Figure 2.8. Münsingen-Rain data. HOMALS solution if absence of an artifact is treated as passive missing data.

other persons. Suppose that the three categories can be labeled "agree," "don't know," and "disagree." Imagine that there is a subset of items with

TABLE 2.8 Artificial Example With Perfect Parallelogram Structure

1	1	1			
	1	1	1	1	
			1	1	1

which some persons agree, whereas other persons either disagree with this same set of items or have no opinion. For instance, if the items are political statements, a "leftist" person will agree with a subset of precisely those items with which a "rightist" person will disagree, whereas a "middle of the road" person has no opinion. In such a situation the HOMALS analysis of the reversed indicator matrix may show good results, in the sense that it contains a HOMALS dimension that identifies "controversial" items.

However, analysis of a reversed indicator matrix could be quite futile. As an example, suppose that the items are questions in a multiple-choice examination. Each item has three response categories, *a, b,* and *c.* One of them is the correct answer. A HOMALS analysis of the reversed indicator matrix will just show that there is a subset of items where *a* is the correct answer, another subset with *b* as the correct answer, and a third one with *c* as the correct answer. Such a result is entirely trivial.

Analysis of the reversed indicator matrix becomes more appropriate in a sorting task. For example, subjects are given 50 cards with names of well-known persons, such as sports figures, artists, statesmen, and scientists. These persons may have lived in different periods of history, may have different nationalities, and so on. The task is to sort the cards into a number of piles according to whatever classification criteria the respondent may select. The only restriction is that there be at least three piles and no more than seven.

In such a situation, the researcher does not even know what criteria subjects have in mind. Nevertheless, a HOMALS analysis of the reversed indicator matrix will unravel which names tend to be together in the same piles, and which not.

The general conclusion seems to be that analysis of a reversed indicator matrix can be profitable if data are comparable to those of a free sorting task in which the precise meaning of the categories depends largely upon a subject's private views.

2.10 Horseshoes

2.10.1 Introduction

Table 2.7, about the Münsingen-Rain graves, rather clearly shows that there is one underlying dimension, the age of the graves. This is expressed

TABLE 2.9 Decomposition of Table 2.8 According to HOMALS Dimensions

.3	.6	.6	.6	.6	.3	.5	.5	.5	−.5	−.5	−.5
.4	.8	.8	.8	.8	.4	0	0	0	0	0	0
.3	.6	.6	.6	.6	.3	−.5	−.5	−.5	.5	.5	.5
						.2	−.1	−.1	−.1	−.1	.2
						−.4	.2	.2	.2	.2	−.4
						.2	−.1	−.1	−.1	−.1	.2

in the fact that the unit elements in the table approximate the structure of a parallelogram. Remember that the layout of this table is based on the HOMALS result (with zero elements treated as passive missing data). Without the help of HOMALS it would have been rather difficult to detect that there is a parallelogram structure in the data.

Nevertheless, Figure 2.8 shows that the HOMALS solution is two-dimensional, and that the graves are ordered along a U-shaped line, a "horseshoe." Such a horseshoe is often found when HOMALS is applied to data where one should expect a one-dimensional structure. The problem is to find out where the horseshoe comes from.

2.10.2 Why Horseshoes?

The answer to the question above is not simple. For one thing, there is no unanimity among authors who have published about this issue (see Van Rijckevorsel, 1987, for a survey). Sometimes the horseshoe is considered to be a mathematical artifact. But it also happens that the second dimension (in which objects in the middle are contrasted with the extreme objects on both sides) is interpreted as meaningful. To illustrate, if we were to find that parliamentarians are ordered along a horseshoe that runs from the political left to the political right, the second dimension might be interpreted in terms of the willingness of political parties to join a government coalition. The argument would be that extreme parties, whether at the left or at the right, are mainly "ideological" and opposed against the established political system, whereas parties in the middle tend to be "pragmatic" and are prepared to work together in a loyal coalition.

2.10.3 Horseshoes as a Mathematical Artifact

Nevertheless, the horseshoe is also a mathematical artifact. This is illustrated with the help of the matrix in Table 2.8. One may look at this table as if it refers to a seriation problem, with 3 graves (rows) and 6 archaeological artifacts (columns). The parallelogram structure in the table clearly shows that there is one underlying dimension.

It will be clear also that Table 2.8 does not have rank 1. For a matrix with rank 1 it must be true that rows (and columns) are proportional to each other. The parallelogram structure implies that rows and columns are certainly not proportional. In fact, the matrix in Table 2.8 has rank 3.

The latter is shown in Table 2.9. Here the matrix of Table 2.8 is decomposed as the sum of three matrices of rank 1. The matrix at the left of Table 2.9 gives the "trivial" HOMALS solution. Its rows and columns are proportional. Moreover, its rows and columns have the same sums as those of Table 2.8. The other two matrices in Table 2.9 therefore have row and column sums equal to zero. They therefore are said to be "double centered." They also have rank 1, because their rows and columns are proportional.

These two matrices correspond to the two HOMALS solutions. The first one shows that there is a linear trend for the first object, an inverse linear trend for the third object, and a zero trend for the second. The second HOMALS solution shows that there is a quadratic (inverse U-shaped) trend in the second object, whereas the other two objects have smaller U-shaped trends.

Although the parallelogram structure in Table 2.8 indicates that there is one underlying dimension, it appears to be impossible to cover the data in the matrix in terms of only one HOMALS dimension. Rows of Table 2.8 contain *step functions*. They never can be described in terms of only one polynomial function.

In fact, if we had more objects in Table 2.8 (n rows instead of 3) and also more categories (m columns instead of 5), we would find $(n - 1)$ or $(m - 1)$ HOMALS solutions (whichever of the two numbers is smaller). All these solutions would show algebraic polynomial functions, with increasing degree. The first HOMALS solution will be linear, the second quadratic, the third cubic, and so on. These are precisely the characteristics of HOMALS solutions that reveal that the original data matrix has a one-dimensional parallelogram structure.

3

ANACOR

3.1 Introduction

The ANACOR program is a direct application of HOMALS in situations where there are only two categorical variables. The theory underlying ANACOR is discussed in Section T.4.5, which also contains a straightforward example of ANACOR.

The acronym ANACOR is an abbreviation of *analyse des correspondances.* This French name goes back to the work of Benzécri et al. (1973), whose writings were very influential in France and elsewhere.

3.2 What Data Can Be Analyzed With ANACOR?

The ANACOR program can be applied if there are data about only two categorical variables. This implies that data can be summarized in a two-way frequency table. Rows refer to categories of one of the variables, and columns to those of the other one. A cell of the matrix gives the frequency of how often a row category goes together with a column category.

In practice, the restriction to only two variables is less severe, because two (or more) variables can be treated as interactive. The variables are thus reduced to only one variable. An example appears in Section 1.3.3, where the variables sex (with 2 categories) and profession (with 7 categories) were merged into one interactive variable with $2 \times 7 = 14$ categories. Another example will be given in Section 3.9. The main point is that ANACOR will give exactly the same results as HOMALS applied to only two variables. This will be discussed further in Section 3.6.

Take a simple example: a bivariate table refers to boys, age 18, and gives their heights in the row categories and their weights in the column categories. The categories have an a priori quantification, given by the interval midpoints. The correlation between height and weight can be calculated simply.

On the other hand, suppose that a frequency table for the same boys refers to color of their hair (rows) and color of their eyes (columns). Is there a correlation between these two variables? The answer could be of some interest to a student of heredity. However, there is no a priori quantification. One might assign some arbitrary a priori quantification to the categories, and calculate the correlation on that basis, but this strategy is rather meaningless and it seems as if in this way one could obtain any correlation between +1 and −1.

However, the latter is not true. The data in the frequency matrix impose restrictions upon the value of the correlation that can be found. In particular, there is an upper bound (and also a lower bound). The question becomes whether categories can be quantified in such a way that the resulting correlation is maximized.

The preceding example is not entirely fanciful. Maung (1941) was among the first to demonstrate this method, and his example was about hair color and eye color among Scottish schoolchildren.

3.3 First ANACOR Example: Marriages in Suriname

Data for the first example are taken from Speckmann (1965). This example has been chosen because the results of ANACOR with these data can be very easily understood. In fact, the results simply confirm what

TABLE 3.1 Marriages in Suriname: ANACOR Solution

A: Frequency Table

		Females				
		c	m	s	a	Sum
Males	C	17	1	4	3	25
	M	1	66	4	2	73
	S	5	4	96	14	119
	A	4	2	18	23	47
	Sum	27	73	122	42	264

B: Category Quantifications Normalized at n = 264

	Dim. 1	Dim. 2	Dim. 3		Dim. 1	Dim. 2	Dim. 3
C	−.59	2.96	.69	c	−.59	2.83	.64
M	1.61	−.02	.03	m	1.62	−.02	.03
S	−.64	−.62	.66	s	−.64	−.63	.61
A	−.58	.03	−2.07	a	−.57	.03	−2.23

C: Category Quantifications Normalized at nr_i^2

C	−.51	1.83	.27	c	−.51	1.75	.25
M	1.40	−.01	.01	m	1.40	−.01	.01
S	−.55	−.38	.25	s	−.55	−.39	.23
A	−.51	.02	−.80	a	−.50	.02	−.86
r_i	.8677	.6182	.3878				

anybody could conclude immediately on the basis of visual inspection of the table.

The data, given in Table 3.1, refer to 264 marriages in Suriname. They categorize husbands and wives in terms of four religious groups: Christian (C), Moslem (M), Sanatin Dharm (S), and Arya Samaj (A) (S and A are two Hindustan religious groups). Table 3.1 is a very simple table in the sense that larger frequencies are mainly on the diagonal, indicating that husband and wife have the same religion. Also, the table is almost symmetric, which predicts that the quantification of husbands (rows) will be about the same as that of wives (columns).

There are three dimensions for the ANACOR solution. In general, for an $r \times c$ frequency matrix there will be $(r − 1)$ or $(c − 1)$ dimensions, whichever number is smallest. The quantifications are given in Table 3.1B, where they are standardized to a sum of squares equal to $n = 264$. Figure 3.1A gives the corresponding graph for the first two dimensions, husbands labeled with capitals, wives lowercase. Because the quantifications of husbands and wives are almost the same, the 8 points in the graph fall into 4 pairs.

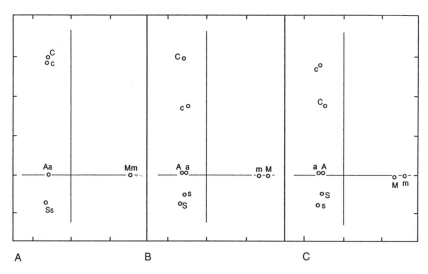

Figure 3.1. Marriages in Suriname (Table 3.1). (A) Scores for husbands and wives have the same normalization. (B) Points for wives are at the center of gravity of points for their husbands. (C) Points for husbands are at the center of gravity of points for their wives.

The bottom line of Table 3.1 shows for each dimension the correlation between the two quantified variables. They are usually called *canonical correlations* (this will be explained in Section 3.6). The first canonical correlation, $r_1 = .8677$, gives a maximum in the sense that it is impossible to find a quantification that produces a higher correlation. The second one, $r_2 = .6182$, stands for a conditional maximum. The condition is that the second quantification of each variable should be uncorrelated with both quantifications on the first dimension. In the same way, the solution on the third dimension must conform to the condition that its quantifications should be uncorrelated with those on the two earlier dimensions. This results in $r_3 = .3878$.

It should be noted, however, that in the example the total number of solutions is restricted to three, because each variable has four categories and it therefore will be impossible to find more than three independent quantifications of them. It follows that once the first two quantifications are determined, only one possible solution for the third quantification remains. Thus the value of r_3 will be fixed, too, in the sense that there is no third solution in which r_3 becomes different from .3878.

For completeness, I should add that the sign of a canonical correlation is changed if the quantification of one of the two variables is given a reversed sign. The canonical correlation therefore has an absolute minimum at $-.8677$. We might say that the squared correlation never can be larger than $(.8677)^2 = .7529$.

In Table 3.1B, the quantifications are standardized to a sum of squares equal to $n = 264$. Table 3.1C gives essentially the same quantifications, with the only difference that they are normalized to a sum of squares equal to nr_s^2 ($s = 1,2,3$). Results in Table 3.1C were used to make the graphs in Figures 3.1B and 3.1C. In Figure 3.1B husbands have coordinates as given in Table 3.1B, and wives as in Table 3.1C. The result is that points for wives appear as the center of gravity of the points for their husbands. In Figure 3.1C the situation is reversed, with points for husbands based on Table 3.1C and those for wives on Table 3.1B. The effect is that points for husbands become the center of gravity of points for their wives. The three graphs in Figure 3.1 are quite similar because the first two canonical correlations are quite large.

The substantive interpretation of the ANACOR results is quite obvious. The first dimension makes a contrast between Moslems and the other three categories. In fact, the frequency table already shows at once that mixed marriages in which one of the partners is Moslem are very rare. The second dimension makes a contrast between C and S. Again, the frequency matrix shows at once that mixed marriages Cs or Sc are comparatively rare, given also that the marginal frequencies of these two religious groups are quite large. The third dimension singles out category A. There are many marriages of type Aa, but (given the marginal frequencies) few marriages of the type As and Sa or of the type Ac and Ca.

3.4 ANACOR and Chi-Square

It is assumed that the reader knows that a chi-square test can be used to find out to what extent rows and columns of a given $r \times c$ frequency matrix are independent. The test takes the following steps:

1. Calculate for each cell the "expected value." This is the product of corresponding row and column total, divided by the total number of observations n. For Table 3.1 the expected values are shown in matrix E of Table 3.2. Rows are proportional to each other and to the row of column totals, similarly for columns. Matrix E has rank 1, in short.

TABLE 3.2 Marriages in Suriname: Chi-Square Solution

E: Expected Values			
2.56	6.91	11.55	3.98
7.47	20.19	33.73	11.61
12.17	32.91	54.99	18.93
4.81	13.00	21.72	7.48

B: Observed Minus Expected Values			
14.44	−5.91	−7.55	−.98
−6.47	45.81	−29.73	−9.61
−7.17	−28.91	41.01	−4.93
−.81	−11.00	−3.72	15.72

C: Values of B^2/E			
81.6	5.1	4.9	.2
5.6	104.0	26.2	8.0
4.2	25.4	30.6	1.3
.1	9.3	.6	32.2

2. Calculate the differences between observed frequencies and expected values. This difference matrix is shown in matrix **B** of Table 3.2.

3. For each cell take the squared difference, and divide by the expected value. This is shown in matrix **C**.

4. Take the sum of the entries in matrix **C**. This sum is called the *chi-square* of the frequency table. In the example, chi-square is equal to 339.35.

A zero value of chi-square means, obviously, that all differences in matrix **B** are equal to zero. Observed frequencies then are equal to the expected values, and the frequency table has rank 1. On the other hand, large value of chi-square means that this "rank 1 hypothesis" must be rejected. The critical value of chi-square depends upon the number of "degrees of freedom," equal to $(r-1)(c-1)$. In the example there are 9 degrees of freedom. Almost any book on statistics contains a table of the critical values of chi-square. The value of 339.35 appears to be highly significant, and it follows that the marriages are not "at random."

Thus far I have done no more than show how chi-square can be calculated. The special point I want to stress, however, is that there is a relation between chi-square and the canonical correlations r_s. This relation can be written as

$$\chi^2 = n \sum r_s^2.$$

For the example, this means

$$339.35 = (264)(.8677^2 + .6182^2 + .3878^2)$$

$$= 198.76 + 100.91 + 39.69.$$

In this way chi-square can be partitioned into distinct portions, one for each ANACOR dimension. These portions might even be tested separately, with $(r - s)(c - s) - (r - s - 1)(c - s - 1)$ degrees of freedom for dimension s. In the example this gives 5 degrees of freedom for the first dimension, 3 for the second, and 1 for the third. These degrees of freedom add up to the total number equal to $(r - 1)(c - 1)$. In the example we find that $5 + 3 + 1 = 9$. A test of the portions of chi-square shows that all three of them are significant.

3.5 Further Elaboration of Chi-Square Decomposition

Not only can the total value of chi-square be decomposed into contributions from the separate ANACOR dimensions, it is also possible to decompose the matrix B of differences between observed and expected values. For the numerical example, this decomposition is given in Table 3.3, where $B = B_1 + B_2 + B_3$.

I shall not give an algebraic proof of such a decomposition, but will describe only how solutions for B_s can be obtained. For a cell of B_s, start with the expected value. Multiply by the canonical correlation r_s. Multiply also by the quantifications of the corresponding row and column category on dimension s. To illustrate, take the upper-left cell of B_2 with value 13.23. It is obtained by taking the expected value 2.56, multiplied by $r_2 = .6182$, and multiplied by the two category quantifications 2.96 and 2.83 (given in Table 3.1 for the second dimension).

As a next step, one can calculate matrices W_s, also shown in Table 3.3. Cells of W_s are found by taking the square of the corresponding cell in B_s, divided by the corresponding expected value. In the example, we find for the upper-left cell of W_2 the value $(13.23)^2/2.56 = 68.5$. Cells of W_s can be added, and the sum will correspond to the portion of chi-square on dimension s.

Results in matrices W_s can be used for the interpretation of the ANACOR dimensions. In the example, W_1 shows clearly that the cells in row M and column m (Moslems) are responsible for the discrepancies. In matrix table B_1 this is further specified, in the sense that there are "too

TABLE 3.3   Chi-Square Solution of Table 3.2 Decomposed According to the Three ANACOR Dimensions

B_1				W_1			
.77	−5.70	3.76	1.16	.2	4.7	1.2	.3
−6.21	45.80	−30.24	−9.34	5.2	103.9	27.1	7.5
4.00	−29.45	19.45	6.01	1.3	26.4	6.9	1.9
1.45	−10.66	7.04	2.17	.4	8.7	2.3	.6
					Sum = 198.76		
B_2				W_2			
13.23	−.27	−13.18	.21	68.5	.0	15.0	.0
−.30	.01	.30	.00	.0	.0	.0	.0
−13.15	.27	13.09	−.21	14.2	.0	3.1	.0
.21	.00	−.21	.00	.0	.0	.0	.0
					Sum = 100.91		
B_3				W_3			
.43	.05	1.86	−2.35	.1	.0	.3	1.4
.05	.01	.22	−.28	.0	.0	.0	.0
1.98	.26	8.48	−10.73	.3	.0	1.3	6.1
−2.47	−.32	−10.55	13.35	1.3	.0	5.1	23.8
					Sum = 39.69		

many" M marriages, and "too few" mixed marriages with either M or m. The matrices W_2 and B_2 show that there are "too many" Cc and Ss marriages, and "too few" Cs or Sc. Matrices W_3 and B_3 reveal that there are relatively "too many" Ss and Aa marriages, and "too few" Sa and As.

3.6 ANACOR and Indicator Matrices

Data from a given $r \times c$ frequency matrix also could be displayed in the format of an indicator matrix G. This matrix will contain two sets of columns, a set G_1 with r columns for the row categories of the frequency table, and a set G_2 with c columns for the column categories. Matrix G will have n rows, one for each object. Such a row will have two entries 1 (for the columns of categories that apply to the object) and 0 entries elsewhere.

Clearly, registering data in the format of such an indicator matrix G is not very efficient. The same information is contained in the much more compact frequency table. Matrix G, therefore, is mainly of theoretical importance. It suggests that we have to deal with two sets of variables. The first one is the set of binary row category variables, and the second is that of the column category variables. It has been shown in Section T.2.3 that we thus obtain a problem in the spirit of classical canonical analysis, where we are looking for weighted sums of variables of each set, in such a way that there is optimal correlation between these two weighted sum variables. The optimal correlation is called *canonical correlation.* Formally, therefore, the indicator matrix G shows that the ANACOR problem is identical to the problem in canonical analysis. To spell it out: The quantification of the row categories of the frequency table will correspond to a weighted sum based upon the columns in G_1. And the columns of the frequency table are quantified in agreement with a weighted sum variable based upon the columns in G_2.

However, the ANACOR program does not produce object scores for the n objects. They can be calculated very easily by taking the average of the two category quantifications that apply to an object. Given the frequency matrix in Table 3.1, there are only $4 \times 4 = 16$ different objects and objects within the same cell obtain identical scores. A couple Cm, for instance, will have object score $(-.59 + 1.62)/2 = .52$ on the first dimension (the category quantifications are taken from matrix B in Table 3.1). It is true that such object scores are not standardized, but this can be easily remedied.

The interpretation of ANACOR in terms of an indicator matrix G also implies that ANACOR is a special variety of HOMALS. Matrix G is the same as the indicator matrix in the situation where HOMALS is applied to only two categorical variables. The advantage of the ANACOR program is that it is entirely adapted to this special situation.

3.7 ANACOR Criteria

ANACOR criteria imply that on each dimension the regression between the two quantified variables is strictly linear. This is illustrated in Section T.6.5. Linearization of regression is in line with the principle of reciprocal averaging. ANACOR also fits in with the idea that there is an upper bound for the correlation between two quantified variables. This upper bound is given by the canonical correlation on the first ANACOR dimension.

Maximization of correlation is probably the oldest criterion, developed in the 1930s. At that time it was thought that one should not rely too much upon the value of a correlation calculated on the basis of a priori quantifications. A quite different value may be obtained if one takes some transformation; for instance, if one takes logarithms of the a priori quantification or if one takes square roots. The question was whether one could obtain any desired correlation by taking an appropriate transformation. It then was found that this question must be answered in the negative, in that there is an upper bound for the resulting correlation, whatever transformation one takes.

A practical conclusion from this finding would be that the a priori quantification can be maintained if it produces a correlation close to the upper bound. On the other hand, transformation of the a priori quantification becomes attractive if it results in a larger correlation. However, in this case it may be useful to have a close look at the transformation plots. It may turn out that the optimal transformation is quite close to some mathematical function (such as a logarithmic function, or a polynomial of low degree). In addition, it may happen that the mathematical functions suggested by the transformation plots make a lot of sense from a substantive point of view. If so, one may even argue that small discrepancies between the optimal quantification and the selected exact mathematical transformation are caused only by small errors in the observed data.

3.8 Second ANACOR Example:
Voting Behavior and Degree of Urbanization

This example is based on Table 3.4, a frequency table in which rows stand for political parties and columns characterize the places of residence of voters who favor those parties. The table is based on a count of votes for the Second Chamber in the elections in 1986, reported in CBS (1987). (See the appendix for more details about the Dutch political system.)

The 6×6 table would allow for five ANACOR solutions. Correlations on these five dimensions are quite small, equal to .1905, .1016, .0224, .0162, and .0062, respectively. For the present example we look only at the first two dimensions. Category quantifications are shown in the lower part of Table 3.4.

The results are not very surprising to anyone familiar with Dutch politics. They show that CDA and "small right" are relatively more popular in rural areas, whereas PvdA and "small left" are more popular in the larger cities, and VVD and D'66 in residential towns. The only surprise, perhaps, is that the canonical correlations are so small. In Figure 3.2 parties are plotted as centers of gravity of the areas.

Again, one may raise all sorts of questions about the interpretation of these results. They may depend on various "hidden variables," such as age,

TABLE 3.4 Voting Behavior and Degree of Urbanization, With ANACOR Solution in Two Dimensions

Frequency Table

	Rural	Industrial/Rural	Commuters	Small Town	Medium City	Large City
PvdA	285	620	355	336	548	903
CDA	482	914	460	337	455	516
VVD	186	308	347	168	233	343
D'66	49	102	104	62	91	153
Small left	21	42	36	27	47	110
Small right	60	97	47	46	43	37

Category Quantifications

	Dim. 1	Dim. 2		Dim. 1	Dim. 2
PvdA	−.185	−.072	Rural	1.274	−.242
CDA	.202	−.045	Industrial/rural	.955	−.826
VVD	.007	.198	Commuters	.318	2.305
D'66	−.147	.119	Small town	.067	−.121
Small left	−.394	−.011	Medium city	−.501	−.375
Small right	.330	−.050	Large city	−1.529	−.232

income, or job opportunities. In a way, the results show only that more research is necessary.

3.9 Third ANACOR Example: Shoplifting

This example concerns 33,101 registered cases of shoplifting in 1977-1978. The data, shown in Table 3.5, were reported by the Dutch Bureau of Statistics (CBS, 1978, 1979) and are also discussed in Israëls (1987). In such data there is, as usual, some ambiguity. First, they depend on cases of shoplifting reported to the police, even though many cases probably occur that are not reported. Second, whether the police are notified may depend to a large extent upon the age and/or sex of the offender, as well as upon the value of the article the person attempted to steal.

Rows of Table 3.5 characterize the nature of the articles shoplifters tried to steal. Columns refer to age and sex of the offenders, treated interactively as if the $9 \times 2 = 18$ categories belong to just one variable. Numerical ANACOR results are not reported here, apart from the first two canonical correlations equal to .591 and .346, respectively.

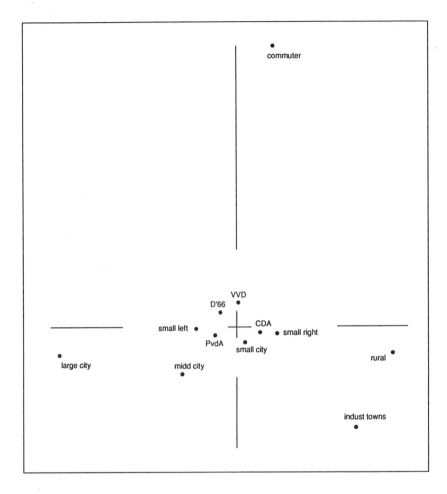

Figure 3.2. Voting behavior and urbanization (Table 3.4). Political parties are plotted as centers of gravity of the six types of districts.

Figure 3.3 gives a graph of the results, with row and column quantifications normalized to sum of squares equal to nr_s^2 ($s = 1, 2$). The figure allows for the following observations:

1. The two broken lines connecting age groups for men and women are, roughly speaking, parallel. This indicates that there is relatively little interaction between age and sex.

TABLE 3.5 Shoplifting

Age	−12	13	16	19	25	35	45	57	65+	Total
Males										
Clothing	81	138	304	384	942	359	178	137	45	2,568
Accessories	66	204	193	149	297	109	53	68	28	1,167
Tobacco										
and food	150	340	229	151	313	136	121	171	145	1,756
Stationery	667	1,409	527	84	92	36	36	37	17	2,905
Books	67	259	258	146	251	96	48	56	41	1,222
Records	24	272	368	141	167	67	29	27	7	1,102
Household										
articles	47	117	98	61	193	75	50	55	29	725
Sweets	430	637	246	40	30	11	5	17	28	1,444
Games	743	684	116	13	16	16	6	3	8	1,605
Trinkets	132	408	298	71	130	31	14	11	10	1,105
Perfumes	32	57	61	52	111	54	41	50	28	486
Hobby										
articles	197	547	402	138	280	200	152	211	111	2,238
Other	209	550	454	252	624	195	88	90	34	2,496
Totals	2,845	5,622	3,554	1,682	3,446	1,385	821	933	531	20,819
Females										
Clothing	71	241	477	436	1,180	1,009	517	488	173	4,592
Accessories	19	98	114	108	207	165	102	127	64	1,004
Tobacco										
and food	59	111	58	76	132	121	93	214	215	1,079
Stationery	224	346	91	18	30	27	23	27	13	799
Books	19	60	50	32	61	43	31	57	44	397
Records	7	32	27	12	12	9	7	13	0	119
Household										
articles	22	29	41	32	65	74	51	79	39	432
Sweets	137	240	80	12	16	14	10	23	42	574
Games	113	98	14	10	21	31	8	17	6	318
Trinkets	162	548	303	74	100	48	22	26	12	1,295
Perfumes	70	178	141	70	104	81	46	69	41	800
Hobby										
articles	15	29	9	14	30	36	24	35	11	203
Other	24	58	72	67	157	107	66	64	55	670
Totals	942	2,068	1,477	961	2,115	1,765	1,000	1,239	715	12,282

2. The two age curves clearly show a break at age 21-30. This is related to the kind of articles stolen. Younger offenders focus on items such as toys, sweets, stationery, and trinkets.

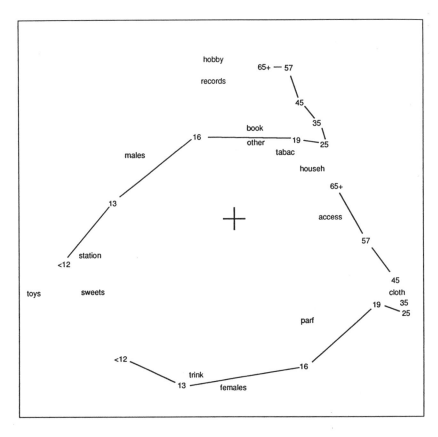

Figure 3.3. Shoplifting (Table 3.5).

3. Especially for young offenders there are differences between the sexes in items stolen. Boys tend to steal toys and sweets, whereas girls are more interested in perfume and trinkets.

4. For the older groups sex differences appear, but the items change: Males want tools, records, and tobacco. Females want household articles, accessories, and clothes.

4 PRINCALS

4.1 Introduction

The acronym PRINCALS comes from *princ*ipal components analysis and *a*lternating *l*east *s*quares. The basic difference between PRINCALS and HOMALS is that the HOMALS program accepts multiple nominal variables only. PRINCALS, however, also accepts the instruction that some or all variables must be treated differently, as numerical, as ordinal, or as single nominal.

These options make PRINCALS a very flexible program. If the user wants all variables to be treated as multiple nominal, PRINCALS results will be the same as in HOMALS. If it is required that all variables be treated as numerical, PRINCALS will produce the same solution as classical principal components analysis. But PRINCALS includes many more possibilities.

4.2 PRINCALS Options

4.2.1 Points of Departure

PRINCALS options are based upon two points of departure. The first is that the user can prescribe whether each individual variable should be

TABLE 4.1 PRINCALS Options

	Single	*Multiple*
Numerical	a	—
Ordinal	b	—
Nominal	c	d

treated as nominal, ordinal, or numerical. The second is that the user may require single or multiple quantification.

4.2.2 Survey of PRINCALS Options

In principle, the two points of departure allow for $3 \times 2 = 6$ different ways of treating each individual variable. These are summarized in Table 4.1, which also shows that there are in fact only four such options.

First, it is impossible to require multiple numerical treatment. The reason is simply that numerical treatment of a variable requires that the quantification of its categories must be the same as the standardized a priori quantification. This implies that multiple numerical quantification contains incompatible requirements. Second, Table 4.1 excludes multiple ordinal treatment. It is not that this requirement is contradictory in itself; there is only a practical reason. It is that the different quantifications of the variable then must have the same order as the a priori quantification. As a consequence, the different quantifications will be highly intercorrelated. Moreover, they also will be highly correlated with the quantification based on single ordinal treatment of the variable. It follows that the multiple ordinal solution therefore does not seem to have much to offer for itself. It is for this reason that the combination multiple/ordinal has not been implemented in the program.

What remains are the four options listed in Table 4.1: single numerical, multiple nominal, single nominal, and single ordinal.

Single numerical. If all variables are treated in this way, the PRINCALS solution will be the same as a PCA solution based upon the a priori quantification. The solution is nested in that the solution for the first p dimensions remains the same when a solution in more than p dimensions is asked for.

Multiple nominal. If all variables are treated as multiple nominal, the PRINCALS solution will be the same as that with HOMALS. The solution is nested.

Single nominal. The effect of this treatment of one or more variables is
that the solution is no longer nested. For instance take a PRINCALS solution in two dimensions. It maximizes the sum of the first two eigenvalues.
But this does not imply that its first eigenvalue will be the same as that in
a PRINCALS solution for only one dimension. It may in fact be smaller.
But this loss then is compensated by a gain in the second eigenvalue.

One might argue that single nominal treatment of a variable makes little
sense. A researcher will treat a variable as nominal if he or she has no a
priori idea of how categories should be quantified. Why then should the
researcher require the same quantification on more dimensions? The
answer to this question is, most likely, that the researcher expects that the
single nominal quantification will be "meaningful." But this implies that
he or she does have some a priori notions about the quantification. If so,
the researcher may better take an a priori quantification that is in agreement with these notions, and treat the variable as single ordinal.

However, it may happen that the researcher has an a priori idea of the
order of only some, not all, categories. The researcher then may hope that
single nominal treatment of the variable will satisfy the expected order of
those special categories, and will assign a rank-order position to the other
categories about which he or she feels uncertain.

It should also be noted that it cannot be proven that a single nominal
PRINCALS solution is unique. It may happen that there is a quantification
A that maximizes the sum of the first p eigenvalues, whereas the same
maximum is obtained when a different quantification B is taken. In fact,
it is possible to construe artificial examples in which this happens. Whether
the PRINCALS program then converges to solution A or to solution B will
depend on the chosen a priori quantification.

This special case shows that the a priori quantification of variables to
be treated as single nominal is not always entirely irrelevant. In actual
practice, however, the dilemma is less disturbing. If a researcher selects
an a priori quantification that favors solution A, he or she will simply
obtain a result that agrees with the a priori notions.

Single ordinal. Again, the PRINCALS solution is not nested once one
or more variables are treated as single ordinal. Results will almost invari-
ably show that the single ordinal treatment of a variable produces a quantification in which some adjacent categories are merged. The reason is
that the PRINCALS program will start with a first guess in which the variable is treated as nominal. If it happens that this solution is ordinal, no
further correction is needed. However, if in the nominal solution some
categories are in the wrong order, a correction can be made by merging

those categories. In other words, nominal treatment of a variable accepts a transformation plot in which there are zigzags. In an ordinal solution such zigzags are forbidden. The easiest way to get rid of them is to flatten the transformation curve by merging categories that form a zigzag. Figure 5.3, in Section 5.8, gives an illustration in which single ordinal and single nominal solutions are compared. The figure shows that the ordinal solution always flattens the irregularities in the nominal solution by merging adjacent categories.

In sum, PRINCALS allows for each individual variable to be treated as numerical, single ordinal, single nominal, or multiple nominal. Once one or more variables are treated as single ordinal or as single nominal, the PRINCALS solution will not be nested, and results will depend on the number of dimensions the user asks for.

4.3 Terminology and Geometric Details

4.3.1 Introduction

The following sections discuss the terminology used in PRINCALS, with the help of an artificial example. This example also serves to illustrate geometric details. The discussion is based largely on Chapters 3 (on PCA) and 4 (on optimal quantification) in the companion volume on theory.

4.3.2 Example

The example is given in Table 4.2. It has been used earlier by Guttman (1968) and by Lingoes (1968), to illustrate their method of multiple space analysis (MSA). It also occurs in GIFI (1983, 1990), where the example is discussed more fully. Here we take it as a purely formal example, with 7 rows (objects) and 5 variables (columns) with 4, 4, 4, 2, and 3 categories, respectively. All variables will be treated as single ordinal.

4.3.3 PRINCALS Results for the First Variable

Results for variable 1, in two dimensions, are given in Table 4.3. This table shows for each of the four categories the corresponding frequency and the single quantification. The latter has the same order as the a priori numbering (highest quantification for category 1, lowest for category 4).

Coordinates of SC points are found by multiplying the category quantification with component loading a_{1s} ($s = 1,2$). Since category quantifications

TABLE 4.2 Data for PRINCALS Example in Tables 4.3 and 4.4

			Variables		
	1	1	1	2	2
	2	2	2	2	2
	1	1	2	1	1
Objects	4	2	4	2	3
	4	4	4	2	3
	3	3	3	1	2
	2	3	3	2	2

have unit variance, coordinates of SC points have variance a_{1s}^2. In agreement with usage in HOMALS, this squared component loading could be called the *discrimination measure*. In PRINCALS, however, it is called the *fit* (per variable, per dimension). Added over dimensions, their sum is called *fit per variable*, in the example equal to .991. Obviously, this fit per variable never can be larger than the number of dimensions in the solution (in the example equal to 2).

TABLE 4.3 PRINCALS Ordinal Solution for First Variable of Table 4.2

Category	Frequency	Category Quantification	Single Dim. 1	Single Dim. 2	Multiple Dim. 1	Multiple Dim. 2
1	2	1.525	1.504	−.207	1.490	−.308
2	2	−.342	−.337	.046	−.373	−.213
3	1	−.381	−.376	.052	−.185	1.434
4	2	−.992	−.979	.135	−1.025	−.197
Sum squares		7.000	6.809	.129	6.854	2.414
Mean sum squares		1.000	.973	.018	.979	.345
Fit			.991		1.324	
Component loadings			.986	−.136		

	Object Scores Dim. 1	Object Scores Dim. 2	Distribution Single Dim. 1	Distribution Single Dim. 2	Distribution Multiple Dim. 1	Distribution Multiple Dim. 2
	1.261	−1.800	−.243	−1.593	−.230	−1.493
	−.259	−.465	.078	−.511	.114	−.252
	1.720	1.185	.216	1.392	.230	1.493
	−.941	−.234	.038	−.369	.084	−.038
	−1.109	−.159	−.130	−.294	−.084	.039
	−.185	1.434	.191	1.382	.000	.000
	−.487	.039	−.150	−.007	−.114	.252
Sum squares	7.000	7.000	.189	6.869	.145	4.585
Mean squares	1.000	1.000	.027	.982	.021	.655
Loss			1.009		.676	

The lower part of Table 4.3 shows the object scores, with objects listed in the same order as in Table 4.2. The crux of the solution is that object points are as much as possible close to their SC points. Table 4.3 therefore also gives the differences (distances) between object points and their SC points. Their sum of squares is minimized over the two dimensions. Per dimension the average sum of squares will be equal to $(1 - a_{1s}^2)$. This value is called the *loss per variable per dimension*. Loss and fit per variable and per dimension add up to 1.

Added over dimensions, the sum of the losses is called *loss per variable,* in the example equal to 1.009. Loss and fit per variable add up to the number of dimensions of the solution. In the example, 1.009 + .991 = 2. The better the fit, the smaller the loss. Equivalently, the larger the spread of SC points around the origin, the smaller their distances to the corresponding object points.

Table 4.3 has more to offer. In particular, it shows coordinates of MC points. They are defined, as usual, as the center of gravity of objects within the same category. For instance, there are two objects in category 1, with scores 1.261 and 1.720 on the first dimension. Their average score is 1.490, and this becomes the first coordinate of the MC point of category 1.

The SC points in a graph will be located on a straight line. The reason is that their coordinates are proportional to the category quantifications, and therefore also proportional to each other. On the other hand, MC points will not necessarily be located on a straight line. In fact, if a variable is treated as single, the PRINCALS solution is not at all interested in the MC points; in the criterion they play no role. Nevertheless, one can calculate the average square of their coordinates, per dimension. In PRINCALS this value is called *multiple fit per variable per dimension.* In the example we find value .979 on dimension 1, and .345 on dimension 2. Their sum, equal to 1.324, is called *multiple fit per variable.* Its value never can be smaller than single fit per variable, because object points are closer to their center of gravity than to any other point, such as their SC point.

Table 4.3 also shows the differences between coordinates of object points and those of corresponding MC points. Their average square per dimension is called *multiple loss per variable per dimension.* For each dimension multiple loss and fit add up to 1. The sum of the losses per dimension is called *multiple loss per variable,* in the example equal to .676. Multiple loss and fit per variable add up to the number of dimensions. In the example, 1.324 + .676 = 2. Multiple loss per variable never can be larger than single loss, for the same reason multiple fit cannot be smaller than single fit.

Results can be graphed, as shown in Figure 4.1. Here we find the seven object points (labeled with their a priori category numbers), their centers

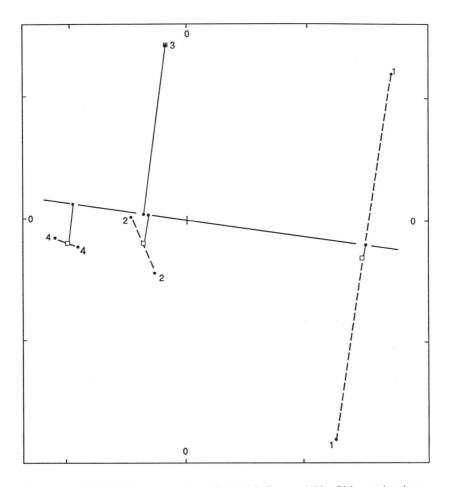

Figure 4.1. PRINCALS example from Table 4.3, first variable. Object points have labels 1, 2, 3, 4, according to their category. MC points are drawn as small squares. Their projections on the straight line are the SC points.

of gravity MC (shown as small squares), and their SC points (located on a straight line). The figure shows that objects are in fact closer to their MC points than to their SC points.

Nevertheless, the PRINCALS criterion for a variable treated as single requires only that objects should be close to their SC points. This has the immediate implication that MC points cannot be far away from corresponding SC points.

In PRINCALS literature the average squared distance between MC and SC points is called *single loss*. The term is confusing, because it may seem to refer to distances between objects and their SC points. For this reason, I prefer the term *relative loss*. In the example, relative loss is equal to 1.009 − .676 = .333. Relative loss must also be equal to the difference between multiple and single fit: 1.324 − .991 = .333. Relative loss may be interpreted as the decrease of fit if MC points are replaced by SC points required to be on a straight line. In other words, if relative loss is small, little will be gained by treating the variable as multiple nominal instead of single ordinal.

Nevertheless, keep in mind that single ordinal treatment of a variable essentially ignores multiple fit and loss. The criterion is concerned only about the SC points and does not bother about MC points. On the other hand, suppose that in PRINCALS a variable is treated as multiple nominal. Now the solution will take MC points into account and will ignore SC points. Actually, in this case the PRINCALS program does not even tell the user where the SC points are located, nor does it give a hint about the decrease in fit to be expected if the variable were treated as single instead of multiple nominal.

A final technical remark: In Figure 4.1 it looks as if the SC points are the projections of MC points on the straight line. This will in fact be true in the space of a complete solution with all possible dimensions (the reason for this is indicated in Section 1.3.6). However, if right angles in the full space are projected on a two-dimensional subspace, they may become distorted and may appear as acute or obtuse angles.

In Figure 4.1, however, no distortion is visible. It looks as if a line connecting an MC point with its SC point still makes a right angle with the straight line through the SC points. The explanation is that in the example this straight line is almost located in the plane of drawing. It would be located exactly in the plane of drawing if the two component loadings had sum of squares equal to 1. In the example their sum of squares is $(.986)^2 + (-.136)^2 = .991$, very close to 1. This implies that in the full space the straight line makes only a very small angle with the plane of drawing, less than 6°. It follows that right angles with the straight line will suffer very little distortion.

4.3.4 Recapitulation of Fit and Loss in PRINCALS

Table 4.4 gives measures of fit and loss for all five variables in the example of Table 4.2, on the first two PRINCALS dimensions. Results for the first variable are the same as in Table 4.3. In Table 4.4, columns 2 and

TABLE 4.4 PRINCALS Measures for Fit and Loss Based on Data in Table 4.2

1	2	3	4	5	6
	Single		Multiple		Relative
Variables	Fit	Loss	Fit	Loss	Loss
1	.991	1.009	1.324	.676	.333
2	.963	1.037	1.166	.834	.203
3	.951	1.049	1.378	.622	.427
4	.922	1.078	.922	1.078	.000
5	.899	1.101	.960	1.040	.061
Sum	4.726	5.274	5.750	4.250	1.024
Average	.945	1.055	1.150	.850	.205

3 add up to 2 (number of dimensions), and the same is true for columns 4 and 5. Column 6 is found by subtracting column 2 from column 4, or by subtracting column 5 from column 3.

The average of column 2 is called *total single fit,* and that of column 4 *total multiple fit.* Similarly, the average of column 3 is *total single loss,* that of column 5 is *total multiple loss,* and that of column 6 is *total relative loss.*

A special situation exists for variable 4. It is a binary variable with only two categories. It follows that its two MC points must be located on a straight line and that they coincide with SC points. As a consequence, relative loss for variable 4 is zero.

4.4 Multiple or Mixed Treatment of Variables

First, suppose that all variables are treated as multiple nominal. As noted above, in this case PRINCALS results will be the same as those of HOMALS. The PRINCALS program will not give component loadings, nor will it identify SC points or show measures of single fit or relative loss.

Second, take the mixed case, where some variables are treated as single and other variables as multiple. For the single variables, PRINCALS will give coordinates of SC and MC points (although the latter have no bearing upon the solution). For the multiple variables, PRINCALS shows only MC points and multiple fit per variable. Total fit is calculated as the average of single fit for single variables, and multiple fit for multiple variables. PRINCALS maximizes total fit for the p dimensions requested. We thus see that the criterion itself becomes a "mixed" one.

How many dimensions are possible for a PRINCALS solution? No
general answer can be given to this question. First of all, if all m variables
are treated as single, there will be m possible PRINCALS dimensions. The
reason is simple. Once the quantification of categories has been decided
upon, PRINCALS becomes identical to classical PCA. On the other hand,
if the jth variable has k_j categories, if all variables are treated as multiple
nominal, and if there are no missing data, the PRINCALS solution will be
the same as that of HOMALS, and the total number of possible dimensions
will be equal to $(\sum k_j - m)$. In the mixed case, the total number of
dimensions will be somewhere between m and $(\sum k_j - m)$.

How many dimensions should be requested? If one requires p dimensions, the PRINCALS solution will maximize the sum of the first p
eigenvalues. Usually, p is taken as rather small, equal to 2 or 3. If all
variables are treated as single, the choice $p = m$ becomes rather absurd,
because in that case the sum of the eigenvalues will be equal to m,
whatever quantification is taken. A choice $p < m$ will produce a solution
that is not nested. Moreover, the solution will maximize the sum of the
first p eigenvalues, and it follows that the sum of the last $(m - p)$ eigenvalues
will be minimized. More specifically, if in this case the sum of the first p
eigenvalues is very close to m, it follows that the sum of the last $(m - p)$
eigenvalues is close to zero, and that the quantified data matrix has rank
close to p. A special example would be the choice $p = (m - 1)$. It minimizes
the last eigenvalue. If it is found that the last eigenvalue is close to zero,
we have a quantification in which linear dependence among the quantified
variables is approximated.

4.5 PRINCALS Example: Suicide Data

4.5.1 Introduction

Data for this example were made available by the Department of Clini-
cal Psychology, University of Leiden (see also Diekstra & Kerkhof, 1982;
Speyer & Diekstra, 1980). The data come from a survey conducted among
517 respondents. The questionnaire contained 12 possible motives for
suicide, and respondents were asked to indicate (on a scale from 1 =
"certainly yes" to 5 = "certainly not") to what degree they considered each
motive to be acceptable. In the present example, only respondents without
missing data are included. The 12 motives are listed in Table 4.5.

For this example all variables were treated as single ordinal. This choice
is based upon two arguments. The first is that response categories clearly

TABLE 4.5 Summary of Variables in Suicide Example

What circumstances might incite you to commit suicide?

1. You are old and decrepit.
2. You have to suffer a lot of pain.
3. You are left alone by your partner.
4. You become severely disabled.
5. You become unemployed.
6. You have a mentally handicapped child.
7. You must be taken to a psychiatric asylum.
8. You cannot beget children.
9. You suffer from an incurable disease.
10. The person dies whom you love best.
11. You cannot find a partner in life.
12. You are guilty of someone's death.

have a meaningful order. The second is that treating the variables as numerical would assume that the a priori quantifications are equidistant. But this assumption is debatable, given the delicate nature of the questions. The second argument is important also for the technical reason that the frequency distributions of the categories deviate very much from the normal distribution. This is shown in Table 4.6. In particular, variables 3, 5, 6, 8, 10, and 11 have extremely skewed distributions, in which category 5 ("certainly not") has larger frequency than all the other four categories taken together.

TABLE 4.6 Category Frequencies for Variables in Table 4.5

	Categories				
Variables	*1*	*2*	*3*	*4*	*5*
1	6	33	127	142	209
2	37	87	203	102	88
3	2	7	51	145	312
4	12	40	172	146	147
5	1	1	2	37	476
6	0	0	5	51	461
7	10	20	111	185	191
8	0	0	8	30	479
9	23	76	197	126	95
10	0	6	43	156	312
11	0	4	16	55	442
12	7	22	148	165	175

4.5.2 PRINCALS Results

For the single ordinal PRINCALS solution $p = 2$ dimensions were asked for. Results are shown in Table 4.7. What immediately strikes the eye is that the distance between the quantifications of categories 4 and 5 is much larger than that between any two other adjacent categories. It is clear that for the respondents category 5 ("certainly not") is quite distinct from all other categories.

It often happens that an ordinal PRINCALS solution merges adjacent categories. In the example we see that in variables 3 and 5, the three categories 1, 2, and 3 obtain the same quantification and therefore are merged. The two variables 8 and 11 are quantified as if they were binary variables (but it is also true that some categories of these two variables have zero frequency). Categories with low marginal frequency especially tend to become merged. In variable 5, for instance, categories 1, 2, and 3 have very small frequency.

This tendency of PRINCALS to merge adjacent categories has been noted already in Section 4.2.2, where it was argued that PRINCALS, as a first step, takes a nominal quantification. But if this intermediate transformation contains zigzags (which do not agree with a monotonic transformation), an ordinal transformation will be approximated by flattening out the zigzags by merging adjacent categories. Moreover, the nominal solution will tend to exaggerate zigzags especially for categories with low frequency. The consequence is that the ordinal solution will show merging, especially for such low-frequency categories.

Table 4.8 gives measures of fit and loss, in the same format as Table 4.4. The values for total fit on the two dimensions are .362 and .165. The PRINCALS solution maximizes their sum, equal to .527.

Because there are no missing data in the example, component loadings are equal to the square root of single fit per variable and per dimension, as shown in Table 4.9. However, given a measure for single fit, we do not know whether the positive or the negative square root should be taken. For the first dimension this poses no problem: One should take the negative root. Why the negative root? The answer is simple and trivial. Take a 5-point response scale, for instance. Usually it will be assumed that objects in category 1 are the highest on the underlying dimension, and objects in category 5 the lowest. On the PRINCALS dimension, therefore, objects in category 1 will obtain high quantification, those in category 5 low. This means that the final quantification reverses the order of the object scores compared with the order of the category quantifications per variable. And so, if one calculates correlations, one will find negative values. This is no more than a built-in feature of the PRINCALS program, and it has no special significance.

PRINCALS

71

TABLE 4.7 Suicide Example: Category Quantifications for the Ordinal PRIN-
CALS Solution in Two Dimensions, and Coordinates of Corresponding SC and MC Points

				Single		Multiple	
Variables	Category	Frequency	Quantification	Dim. 1	Dim. 2	Dim. 1	Dim. 2
	1	6	−1.88	1.19	−.68	.84	−1.29
	2	33	−1.50	.95	−.54	.80	−.82
1	3	127	−.90	.57	−.33	.54	−.38
	4	142	−.48	.31	−.18	.43	.04
	5	209	1.17	−.74	.42	−.77	.37
	1	37	−1.22	.79	−.65	.69	−.78
	2	87	−.98	.64	−.52	.59	−.58
2	3	203	−.35	.23	−.18	.21	−.20
	4	102	.24	−.16	.13	.01	.34
	5	88	1.99	−1.30	1.06	−1.38	.97
	1	2	−1.30	.79	.33	−.41	−.09
	2	7	−1.30	.79	.33	1.13	.44
3	3	51	−1.30	.79	.33	.87	.13
	4	145	−1.21	.74	.30	.71	.37
	5	312	.81	−.50	−.20	−.49	−.21
	1	12	−1.81	1.26	−.70	1.16	−.89
	2	40	−1.25	.87	−.48	.81	−.58
4	3	172	−.68	.47	−.26	.44	−.32
	4	146	−.21	.15	−.08	.24	.07
	5	147	1.50	−1.04	.58	−1.07	.53
	1	1	−3.76	1.83	2.01	.72	1.73
	2	1	−3.76	1.83	2.01	1.99	.34
5	3	2	−3.76	1.83	2.01	2.33	2.96
	4	37	−3.37	1.64	1.80	1.64	1.80
	5	476	.29	−.14	−.16	−.14	−.16
	1	0	—	—	—	—	—
	2	0	—	—	—	—	—
6	3	5	−3.58	1.97	1.98	1.95	1.99
	4	51	−2.79	1.54	1.54	1.54	1.54
	5	461	.35	−.19	−.19	−.19	−.19
	1	10	−2.19	1.34	−.28	1.33	−.32
	2	20	−1.56	.95	−.20	.93	−.31
7	3	111	−.67	.41	−.09	.39	−.18
	4	185	−.61	.37	−.08	.39	.00
	5	191	1.26	−.77	.16	−.77	.15

continued

TABLE 4.7 Continued

				Single		Multiple	
Variables	Category	Frequency	Quantification	Dim. 1	Dim. 2	Dim. 1	Dim. 2
	1	0	—	—	—	—	—
	2	0	—	—	—	—	—
8	3	8	−3.55	1.55	2.00	1.50	2.07
	4	30	−3.55	1.55	2.00	1.50	2.07
	5	479	.28	−.12	−.16	−.12	−.16
	1	23	−1.34	.90	−.65	.62	−1.03
	2	76	−1.09	.73	−.53	.70	−.57
9	3	197	−.46	.31	−.22	.31	−.21
	4	126	.19	−.13	.09	−.01	.25
	5	95	1.89	−1.27	.92	−1.34	.82
	1	0	—	—	—	—	—
	2	6	−1.53	.99	.25	.92	.55
10	3	43	−1.51	.98	.25	1.02	.08
	4	156	−1.14	.74	.19	.73	.23
	5	312	.81	−.52	−.13	−.52	−.14
	1	0	—	—	—	—	—
	2	4	−2.43	1.29	.98	1.21	.27
11	3	16	−2.43	1.29	.98	1.26	.59
	4	55	−2.43	1.29	.98	1.30	1.15
	5	442	.41	−.22	−.17	−.22	−.17
	1	7	−2.21	1.40	−.13	1.45	.38
	2	22	−1.47	.94	−.08	.91	−.37
12	3	148	−.93	.59	−.05	.60	.02
	4	165	−.26	.16	−.01	.16	−.10
	5	175	1.30	−.83	.08	−.83	.10

Therefore, in Table 4.9 one might as well reverse the signs. We then obtain Figure 4.2 (in which the directions of axes x_1 and x_2 are reversed). The figure shows that on the second dimensions the variables are ordered 8, 6, 5, 11, 3, 10 on one side, and 12, 7, 1, 4, 8, 2 on the other. This suggests that there is a distinction between "social motives" on one side and "physical motives" on the other. This distinction goes together with a distinction that can be made in Table 4.6, that "social motives" have a very skewed distribution (a great majority of the respondents reject these motives), whereas "physical motives" have a much more even distribution.

Figure 4.2 strongly suggests that the first PRINCALS dimension distinguishes between objects who reject suicide whatever the motive might be and objects who do not reject suicide under all circumstances. The second

TABLE 4.8 Suicide Example: PRINCALS Measures of Fit

Variables	Single		Multiple	
1	.400	.131	.409	.155
2	.426	.284	.434	.296
3	.373	.063	.381	.069
4	.483	.150	.486	.160
5	.237	.286	.240	.295
6	.302	.304	.302	.304
7	.373	.016	.373	.021
8	.191	.319	.192	.320
9	.451	.235	.458	.249
10	.422	.028	.422	.031
11	.282	.164	.282	.175
12	.406	.003	.406	.014
Mean	.362	.165	.365	.174
Total fit	.527		.539	
Relative loss		.012		

dimension indicates that respondents who reject suicide for social motives may accept the validity of physical motives (and vice versa).

4.5.3 Background Variables

A questionnaire with items such as those described above will also collect background information about the respondents—age, religion, and so on. These data should be treated as passive variables. It stands to reason that respondents with strong religious convictions will be inclined to reject

TABLE 4.9 Suicide Example: Component Loadings

1	−.633	.362
2	−.653	.533
3	−.611	−.251
4	−.695	.387
5	−.486	−.535
6	−.550	−.551
7	−.611	.128
8	−.437	−.565
9	−.671	.485
10	−.649	−.166
11	−.531	−.405
12	−.637	.057

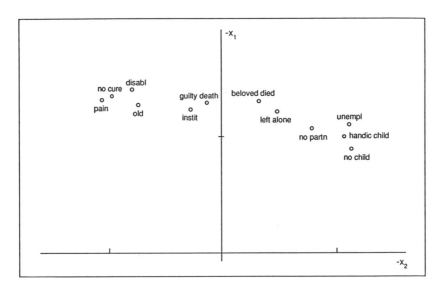

Figure 4.2. Component loadings for suicide data (Table 4.9), with directions of the axes reversed.

suicide under all circumstances. One may also expect that older respondents are more aware of the pertinence of physical motives, whereas younger respondents have more understanding of the relevance of social motives. Such speculations can be checked even if background variables are treated as passive. In Figure 4.2 one then could plot MC points of categories of background variables. The substantive interpretation of the dimensions in the figure then would become more solid, more convincing.

4.5.4 MC Solutions

In the example above, variables are treated as single ordinal, but the multiple fit of each variable is hardly better than the single fit. This indicates that treatment of the variables as multiple nominal will produce about the same results.

I shall conclude this chapter with some remarks about the three variables for which results are shown in Figure 4.3. Figure 4.3A refers to variable 2 (pain). In Figure 4.2 this variable has largest loading on the second dimension—pain is a physical motive. Figure 4.3C refers to variable 8 (no children), which is in Figure 4.2 on the opposite side (social motive).

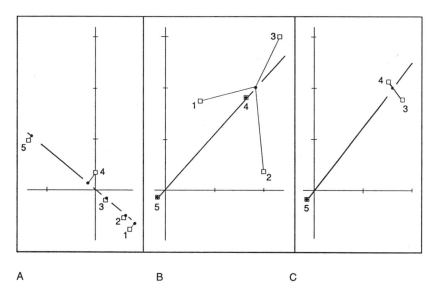

A   B   C

Figure 4.3. Suicide data. (A) For variable "pain." (B) For variable "unemployed."
(C) For variable "no procreation."

Figures 4.3A and 4.3C show that in both cases MC points are quite close
to SC points. Figure 4.3B refers to variable 5 (unemployment), also a
social motive, according to Figure 4.2. Figure 4.3B shows that for this
variable, the categories 1, 2, and 3 are merged in the PRINCALS single
quantification, although their MC points are quite different. The reason
for this discrepancy is probably that the three categories have very small
marginal frequency.

A tentative conclusion is that the PRINCALS solution, with all variables
treated as single ordinal, may not be very much different from a HOMALS
solution where all variables are treated as multiple nominal. The major
difference between these two solutions will be that HOMALS tends to
exaggerate differences between categories with low frequency. This risk
is always present in a HOMALS solution. Treatment of the variables as
single ordinal implies that such a risk is counterbalanced, in the sense that
low-frequency categories will be merged.

5

CANALS

5.1 Introduction

CANALS is the "nonlinear" variety of classical canonical analysis (CA), introduced in Section T.2.3. In that section it was explained that CA has to do with two sets of variables. For each of the two sets, we want to form a weighted sum (called *canonical variate*). The criterion is to select the weights in such a way that there is optimal correlation between the two canonical variates. These correlations are called *canonical correlations*. Correlations between canonical variates and original variables are called *canonical loadings*. The criterion requires that canonical variates on the second dimension must be uncorrelated with those on the first dimension. Generally, canonical variates for some dimension must be uncorrelated with canonical variates on any other dimension.

It is easy to guess what the objective of CANALS is. Given that there are two sets of categorical variables, how should the categories of these variables be quantified in order to obtain, after quantification, optimal canonical correlations?

The acronym CANALS is derived from *c*anonical *an*alysis by means of *a*lternating *l*east *s*quares.

5.2 Some Assumptions Underlying CA

Before discussing CANALS itself, I would like to make some remarks about the general characteristics of canonical analysis. The first point is that there are two sets of variables. Moreover, this division of variables into two sets must be made on an a priori basis, independent of any analysis of the data. For instance, if in a questionnaire respondents are asked to state their preferences for political parties and also are asked to give their opinions about a number of political issues, then it is clear a priori that questions of the first type form a set that differs from those of the second type.

The second point is that the two sets are treated as symmetrical. It does not matter which set is called the "first" one and which the "second."

A counterexample is that the first set refers to variables measured at some point in time, whereas the second set contains the same variables measured at a later point in time. In this case the relation between the two sets is asymmetric. It seems reasonable enough to try to predict variables in the second set on the basis of results obtained in the first set, but the reverse seems absurd, to "predict" earlier variables from later ones.

The third preliminary point is that the CA solution does not depend on some specific ordering of variables within the sets. For instance, suppose that data in the first set are formed by answers given by husbands to four questions, whereas the second set contains answers of their wives to the same four questions. Clearly there now is a 1-1 correspondence between variables in the first set and those in the second. There are four *pairs* of variables. Canonical analysis ignores entirely that there is such a special match between pairs of variables. CA will treat the data in the same way as when variables in the second set were entirely different from those in the first set.

5.3 Properties of CA

5.3.1 Canonical Correlations

Suppose we have two sets of two variables, and that we take two dimensions for the canonical solution. This solution solves for canonical

TABLE 5.1 Matrices of Correlations Between Canonical Variates

A:	y_{11}	y_{12}	y_{21}	y_{22}	B:	y_{11}	y_{12}	y_{13}	y_{21}	y_{22}
y_{11}	1	0	r_1	0	y_{11}	1	0	0	r_1	0
y_{12}	0	1	0	r_2	y_{12}	0	1	0	0	r_2
					y_{13}	0	0	1	0	0
y_{21}	r_1	0	1	0	y_{21}	r_1	0	0	1	0
y_{22}	0	r_2	0	1	y_{22}	0	r_2	0	0	1

variates on each dimension. They are defined as weighted sums of the variables within the same set. Call them y_{11} and y_{12} for the first set, and y_{21} and y_{22} for the second. Table 5.1A shows what the matrix of correlations between these canonical variates will look like. Within the same set, canonical variates are uncorrelated. The off-diagonal submatrix of correlations between variates from different sets will be a diagonal matrix. The correlations on its diagonal are called *canonical correlations.*

On the other hand, suppose that the first set contains m_1 variables, and the second m_2. Without loss of generality, let $m_1 > m_2$. The matrix of correlations between canonical variates now contains two diagonal blocks. The first one is an $m_1 \times m_1$ identity matrix left-above, the other one an $m_2 \times m_2$ identity matrix right-below. The remaining right-upper off-diagonal submatrix will be a rectangular $m_1 \times m_2$ block. The upper $m_2 \times m_2$ part of this block will be a diagonal matrix, with canonical correlations as its diagonal elements. The remaining $(m_1 - m_2)$ rows of the block contain zeros. This is exemplified in Table 5.1B, for $m_1 = 3$ and $m_2 = 2$. In this example there is, so to speak, a third canonical variate y_{13}, which has no canonical "partner" in the second set. This variate has the property of being uncorrelated with all original variables in the second set.

For completeness, it must be added that it may happen that there are zero canonical correlations. In Table 5.1A or 5.1B the value of r_2 might be equal to zero.

5.3.2 Canonical Weights and Canonical Loadings

Table 5.2 gives a fictitious example of results that could be obtained for the first dimension of a canonical solution. In this example the first set has $m_1 = 2$ variables, the second $m_2 = 3$. There are two canonical variates y_{11} and y_{21}. The table shows the *canonical weights* of these variates. For instance, we have

TABLE 5.2 Fictitious Example of Canonical Weights and Canonical Loadings

		Weight	Component Loadings y_{11}	y_{21}
First set	x_{11}	.872	.971	.728
	x_{12}	.258	.592	.444
Second set	x_{21}	.601	.627	.836
	x_{22}	−.221	−.654	−.872
	x_{23}	.433	.527	.702

$$y_{11} = (.872)x_{11} + (.258)x_{12}.$$

The table also shows *canonical loadings.* They are defined as the correlations between canonical variates and original variables. The purpose of the example is to illustrate some relations between weights and loadings.

First, if original variables within one set were uncorrelated with each other, their weights would be equal to their loadings on the canonical variate of the same set. In practice, however, original variables within a set will be intercorrelated. The consequence is that relations between weights and loadings become much less simple. In Table 5.2, for instance, it can be seen that x_{22} has smallest (negative) weight, but has largest (negative) loading.

Second, without proof, we note that the cross-products of weights and loadings on the canonical variate of the same set add up to 1. In the example, we find for the first set that

$$(.872)(.971) + (.258)(.592) = 1$$

and for the second

$$(.601)(.836) + (−.221)(−.872) + (.433)(.702) = 1.$$

The implication is that it will be impossible for all variables to have high positive weights and, at the same time, high positive loadings, because this would result in a sum of cross-products larger than 1.

In the example in Table 5.2, the first canonical correlation is equal to r = .750. If loadings of original variables on their own canonical variate are multiplied by r, you find the loadings on the other canonical variate. In the example, x_{11} has loading .971 on y_{11} and loading $(.971)(.750) = .728$

on y_{21}, whereas x_{21} has loading .836 on y_{21} and $(.836)(.750) = .627$ on y_{11}. It follows that cross-products of weights and loadings on the canonical variate of the other set must have sum equal to r.

It is important to note that a large canonical correlation does not imply that loadings will be large. For instance, in earlier work, I presented an example where the canonical correlation is equal to 1 but all canonical loadings are close to zero (Van de Geer, 1986). Phrased somewhat differently, one could say that a large canonical correlation does not necessarily imply that the canonical variates "explain" much of the variance of the original variables.

Finally, we have the problem of how a canonical variate should be interpreted. One may look at canonical weights and say that a canonical variate depends mostly on original variables with large positive or negative weight, not on variables with weight close to zero. But one may also look at the loadings and conclude that a canonical variate is related mainly with original variables that have large positive or negative loading, and has little to do with variables that have almost zero loading. It is indicated above that these two approaches may lead to quite different interpretations of a canonical variate, because an original variable may have large weight and small loading, or the other way around.

If an original variable has high loading but small weight, this would mean that the variable does not contribute anything that is not already covered by other variables. Such a variable might be dropped without much change in the canonical variate—the variable is superfluous.

On the other hand, if a variable has large weight but small loading, the reason will probably be that it serves as a "suppressor." For instance, we may have a canonical variable that is unrelated to age and that depends on other variables. But it happens that the other variables in the set are correlated with age. We then need a large weight for variable age in order to counterbalance the unwanted age component in other variables.

5.4 Graphs

To simplify the exposition, let us take the example of a canonical solution in two dimensions, with canonical variates y_{11} and y_{12} for the first set and y_{21} and y_{22} for the second.

To make a graph, we can start with the first set. Object points can be plotted by taking their coordinates from y_{11} and y_{12}. Since these two canonical variates are standardized and uncorrelated, the object points will form a "circular cloud," in the sense that they have unit variance not only in the direction of y_{11} and y_{12}, but in any other direction as well.

As a next step we can represent a variable as a line through the origin and the point defined by the component loadings of the variable on y_{11} and y_{12}. Categories of the variable will have SC points located on this line. Their coordinates are obtained by multiplying the component loadings by the standard score of a category. The same type of graph can be made on the basis of the second set, with y_{21} and y_{22} as axes.

Of course, the four canonical variates span a four-dimensional space. The two graphs represent projections on two planes within that space. It therefore would be another step to represent object points in the second plane (their coordinates are given by y_{21} and y_{22}) by their projections on the first plane. These projections will be points with coordinates $y_{21}r_1$ and $y_{22}r_2$. In this way we obtain in the first graph two points for each object, one of them with coordinates from y_{11} and y_{12}, the other one with coordinates from $y_{21}r_1$ and $y_{22}r_2$. Altogether, we obtain *n pairs* of points. If the canonical solution is a good one, object points within the same pair will be close together. More precisely, in the direction of the first dimension object points within the same pair have an average squared distance of $(1 - r_1^2)$, whereas in the direction of the other dimension this average square is equal to $(1 - r_2^2)$.

In exactly the same way, object points of the first plane can be projected on the second plane. The projections have coordinates $y_{11}r_1$ and $y_{12}r_2$. Again we thus find *n* pairs of points, close to each other if the canonical correlations are large.

5.5 Basic Properties of CANALS

Given two sets of categorical variables, the objective of CANALS is to quantify the categories in such a way that after transformation the canonical solution is optimal. More precisely, the sum of the squared canonical correlations on *p* dimensions will be maximized. The choice of *p* is left to the user.

The following should be noted:

1. CANALS is restricted to *single* quantification. The available options for each individual variable therefore are to treat it as numerical, as single ordinal, or as single nominal.
2. A CANALS solution is *not nested.* The only exception is that all variables are treated as numerical. The CANALS solution then becomes identical to a classical CA solution based upon a priori quantifications, and this solution is nested. Perhaps I should repeat here that a binary variable always will be treated as numerical.

3. Missing data are treated as *multiple active*. CANALS has no option for treating them as passive, but it remains possible to treat them as single active. The user then must use the trick that missing data are read in not as missing, but as if they are in one additional category of their variable.

4. The CANALS criterion is to maximize the sum of squares of the first p canonical correlations. The choice of p is up to the user, but should not be larger than the number of variables in the smallest set.

An alternative criterion would be to maximize not the sum of the squared canonical correlation, but the sum of their absolute values. If $p = 1$ this makes no difference. Also, if all variables are treated as numerical, the two criteria come to the same thing, because in this case the solution will be nested. However, with $p \geq 2$ the two criteria may produce solutions that are slightly different, although I do not know under what circumstances such a divergence may occur.

The alternative criterion is mentioned for a special reason. The program OVERALS (to be discussed in Chapter 6) is applicable when variables are partitioned into two or more sets. One therefore might expect that OVERALS gives the same solution as CANALS when there are just two sets. However, the OVERALS criterion implies that the sum of the absolute values of the canonical correlations will be maximized. OVERALS and CANALS therefore may produce somewhat different results.

5.6 Special Cases

Section T.2.5 mentions some special cases of classical CA; these are also valid for CANALS. Two examples follow.

First, we might want a CANALS solution with some variable partialed out. This can be achieved by including this variable in *both* sets. This will result in a perfect canonical correlation of 1, with canonical variates based upon the duplicated variable. By itself this result is quite trivial and of no interest. However, on subsequent dimensions the canonical variates will be uncorrelated with those of the first dimension. This implies that in the later solutions the duplicated variable is partialed out.

Another special case is discriminant analysis. The basic idea is that data are collected about a number of variables for objects that belong to different subgroups. An example would be that the variables refer to attitudes of persons who can be classified in subgroups on the basis of their preference for one of the existing political parties. The task is to find weighted sums of the attitude variables, so that these weighted sums make a good discrimination between the subgroups. Analysis of the data then can be brought within the framework of canonical analysis if the classification into subgroups is translated into one or more nominal variables. These nominal variables then are taken as one set, and the attitudes form the second set. In this way the problem is brought down to an analysis of relations between the two sets.

CANALS thus performs what in the classical literature is known as *discriminant analysis.* Moreover, the options in CANALS also allow for varieties of nonlinear discriminant analysis. The only problem is that CANALS produces output that is not in the format usual for discriminant analysis. The user, therefore, is left with the job of translating the output so that results can be presented in classical style. Van der Burg (1983) shows how such a translation can be done.

5.7 First CANALS Example: Dutch Parliament

5.7.1 Introduction

Data for this example were taken from a questionnaire distributed among the members of the Second Chamber of Dutch Parliament in 1972. More information about this questionnaire can be found in Daalder and Rusk (1972). The example is also used in GIFI (1990).

Two sets of variables were taken from this questionnaire. The first set consists of data from questions asking the parliamentarians to rank order the political parties represented in the Chamber on the degree to which they felt congenial with them. For the present example these data have been restricted to only four of the political parties: PvdA (39 representatives), KVP (35), VVD (16), and AR (13). The second set of variables refers to the seven political issues listed in Table 5.3. Subjects were asked to express their opinions on a 9-point scale.

The Second Chamber has 150 seats. Data were available from $n = 138$ parliamentarians. In the CANALS analysis all variables were treated as single ordinal, and the number of dimensions was set at $p = 2$.

5.7.2 First Analysis

The CANALS analysis resulted in a first dimension with canonical correlation equal to 1. However, this appears to be an artificial result. It is caused by the fact that one respondent is the only one to express extreme dislike of KVP, whereas this respondent has missing data on one of the issues. In other

TABLE 5.3 Seven Political Issues Used in Parliament Survey Example

DEVELOPMENT AID

The government should spend more money on aid to developing countries.   1 9   The government should spend less money on aid to developing countries.

ABORTION

The government should prohibit abortion.   1 9   A woman has the right to decide for herself.

LAW AND ORDER

The government takes too strong action against public disturbances.   1 9   The government should take stronger action against public disturbances.

INCOME DIFFERENCES

Income differences should remain as they are.   1 9   Income differences should become much less.

PARTICIPATION

Only management should decide important matters.   1 9   Workers too must participate in important decisions.

TAXES

Taxes should be increased for general welfare.   1 9   Taxes should be decreased.

DEFENSE

The government should insist on shrinking the Western armies.   1 9   The government should insist on maintaining strong Western armies.

words, the respondent is unique in one of the categories of the first set, and also in one category of the second. The consequence is that CANALS gives this object an extreme score on both canonical variates, whereas all other 137 respondents are merged and obtain identical scores. It is easy to see that this result is artificial in that it is based upon just one object.

I report this result for two reasons. First, it illustrates that a perfect canonical correlation may appear if there is an object in the position of outlier, in the sense of being unique in some category of each of the two sets. Second, treating missing data as multiple active implies that categories are created, perhaps many of them, that contain only one object. The consequence is that a perfect canonical correlation will appear if there is an object with missing data in variables of both sets. At first sight it seems attractive to treat missing data as multiple active, but this example shows that this treatment has its peculiar risks in CANALS.

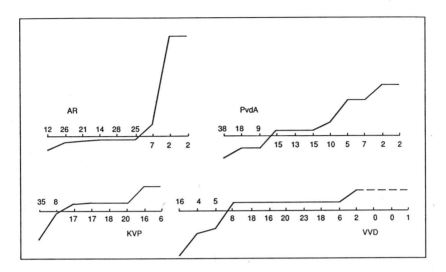

Figure 5.1. Parliament survey. Category quantification of preference variables. Positive quantification (at the right of scales) means antipathy; negative quantification (at the left) means sympathy. Numbers indicate the frequency of each category.

5.7.3 CANALS Results of Second Analysis

After elimination of the outlying object, CANALS was repeated over the remaining $n = 137$ objects. This produced canonical correlations $r_1 = .921$ and $r_2 = .916$.

Category quantifications are shown in Figure 5.1 for preferences, and in Figure 5.2 for issues. The frequencies indicated in these figures do not always add up to $n = 137$ because there are missing data.

In Figure 5.1 (preference) the quantification goes up with antipathy. On the whole, the categories at the left (sympathy) have relatively large frequencies, and the categories at the right (antipathy) have small frequencies.

Figure 5.2 (issues) deserves more comment. As to development aid, the first five categories are more or less merged. A great majority of parliamentarians do not want to reduce developmental aid—only a few of them (12) do. As to income differences, the five highest categories are more or less merged. Most parliamentarians are in favor of a decrease in income differences. The quantification of this issue discriminates mainly between the infrequent categories at the opposite side. Similar remarks could be made for the other issues, but the figure more or less speaks for itself.

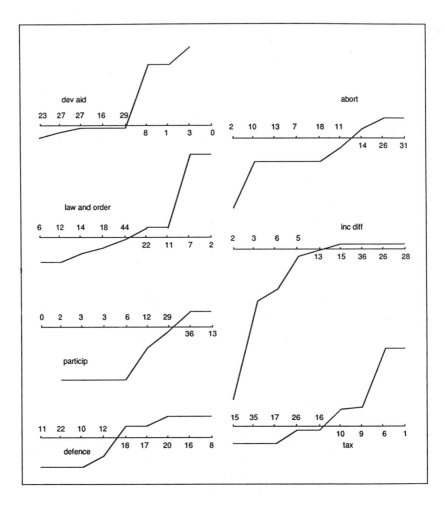

Figure 5.2. Parliament survey. As in Figure 5.1, but now for the issue variables. Categories are ordered from 1 (left) to 9 (right), corresponding to Table 5.3.

Table 5.4 gives canonical loadings. In this table loadings of preferences have been reversed in sign, because it is easier to think of these variables in terms of "sympathy" than of "antipathy." Figure 5.3 gives a graph of the loadings. The graph is surrounded by a circle with unit radius. The reason is that loadings of a variable cannot have sum of squares larger than 1, so that points in the graph must be located in the interior of the unit circle.

TABLE 5.4 Parliament Survey: Loadings on Canonical Variates of First Set (Issues) and Second Set (Preferences)

| | Issues | | Preferences | |
	Dim. 1	Dim. 2	Dim. 1	Dim. 2
Development aid				
less	−.482	−.099	−.445	−.091
Abortion free	−.120	−.586	−.110	−.535
Order stronger	−.429	.495	−.395	.453
Income differences				
smaller	.767	.335	.708	.309
Participation of				
workers	.771	−.232	.710	−.214
Tax decrease	−.717	.258	−.660	.237
Defense maintain	−.466	.546	−.431	.497
Preference for:				
PvdA	.555	−.707	.603	−.772
AR	.635	.472	.690	.516
KVP	.222	.342	.241	.373
VVD	−.830	−.059	−.901	−.064

The figure shows that sympathy for VVD goes together with no decrease in income differences, less development aid, reduction of workers' participation, and rejection of tax increase. Sympathy for PvdA is related to tolerance with respect to law and order, reduction of military expenditure, higher taxes, and more workers' participation. The vertical direction in the graph is dominated by the abortion issue and shows that a ban on abortion tends to go together with sympathy for the denominational parties KVP and AR.

Table 5.5 gives coordinates of the MC points of the 11 political parties from which respondent representatives came. These parties now are treated as categories of a passive variable. Figure 5.4 shows the corresponding graph. The 4 + 7 variables are shown as straight lines. They have the same directions as in Figure 5.3, but in Figure 5.4 their length is irrelevant. The figure also contains two MC points for each of the 11 political parties. They are indicated with a small circle for preferences and a small square for issues. The figure clearly confirms the interpretation suggested above for Figure 5.3. However, one must be familiar with the Dutch political situation in 1972 to appreciate the minor details. Three examples follow:

1. The larger denominational parties, KVP, AR, and CHU, are somewhat in the middle of the political "right-left" dimension. But the two small denominational

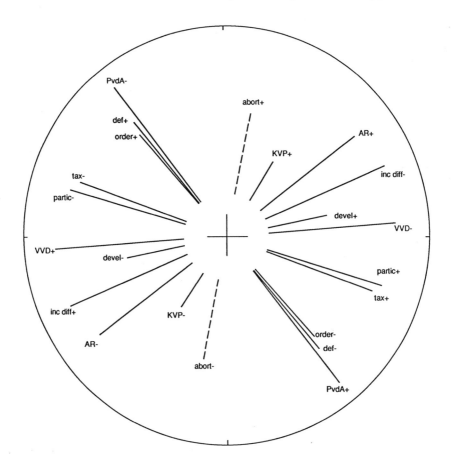

Figure 5.3. Parliament survey. Component loadings (in the interior of a circle with radius 1).

parties, SGP and GPV, are much more to the "right," and also much more outspoken in their rejection of free abortion.

2. The two MC points of DS'70 are far apart. The point for issues is more to the right, and that for preferences more to the left. One is tempted to say that this party, founded in 1970 by "dissidents" from PvdA, does not quite fit in the overall political pattern, and this may explain why DS'70 was dissolved shortly after 1972.

3. PSP is close to PvdA with respect to issues. On the other hand, PSP representatives have on the average little sympathy for PvdA.

TABLE 5.5 Parliament Survey: For Representatives of Each Political Party, Co-ordinates of the Center of Gravity Based on Scores on Two Canonical Variates for Issues and Based on Scores for Preferences

| | *Issues* | | *Preferences* | |
	Dim. 1	*Dim. 2*	*Dim. 1*	*Dim. 2*
PSP	.710	−.966	.068	−.318
PPR	.916	−.410	.623	−.409
PvdA	.545	−.858	.588	−.932
D'66	.155	−.590	.142	−.546
DS'70	−.778	.520	−.345	−.485
AR	.272	.659	.362	.671
KVP	.419	.529	.437	.590
WD	−2.102	−.416	−2.290	−.417
CHU	.110	.679	−.020	.597
GPV	−.863	1.999	−.518	1.643
SGP	−1.159	2.646	−.901	2.348

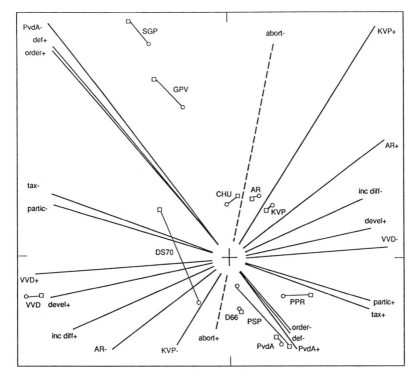

Figure 5.4. Parliament survey. Centers of gravity of the 11 political parties on the canonical variates for preferences (small circle) and for issues (small square).

These three examples illustrate that Figure 5.4 is in many ways an interesting figure. Even where the figure is not a sufficient basis for drawing final conclusions, it may give rise to meaningful questions for further research.

5.8 Second CANALS Example: Crime and Fear

Section 1.3 gives a PRIMALS example with variables related to effectiveness of measures of crime prevention and variables related to feelings of anxiety in various situations (see Table 1.1). It is clear that there are two sets of variables, and it is only natural to attempt a CANALS analysis. Results are displayed in Table 5.6, both for an analysis in which all variables are treated as single ordinal and for an analysis with all variables single nominal. Figure 5.5 gives a graph of quantifications. They confirm a remark made earlier, to the effect that an ordinal solution will come down, mainly, to merging those adjacent categories that in the nominal solution are quantified in the wrong order.

CANALS results on the first dimension are, on the whole, in agreement with those of PRIMALS. They show that absence of fear tends to go together with belief in the efficacy of "social" prevention and disbelief in the efficacy of "punitive" measures. But the canonical correlation is rather small. One may note that variables 2 (prison) and 10 (no help from bystanders) have little impact on the first dimension.

Interpretation of the second dimension is more difficult. The nominal solution suggests a (small) second canonical correlation based mainly upon the variables, 2 and 10, which play no role at the first dimension. From a formal technical point of view the result on the second nominal dimension might be caused by the fact that the nominal quantification of these two variables is U-shaped. The ordinal solution precludes such a quantification, so that the second ordinal dimensions must be different. It is based upon a contrast of variable 1 with variables 3 and 5 in the first set, and of variable 7 with variables 9 and 10 in the second set. In the example this second ordinal dimension seems to be of little importance.

As to the nominal solution, the following general remark can be made. Assume that there are variables of which the categories are running from one extreme position to the opposite one, with such neutral categories as "no opinion" in the middle. I am inclined to believe that a nominal CANALS solution then will often give a U-shaped quantification of the categories of some of those variables, with the effect that it becomes possible to obtain dimensions on which objects with "no opinion" are contrasted with objects in the more extreme categories, no matter on which side of the variable. Such a dimension discriminates between respondents

TABLE 5.6 Crime and Fear Example: Loadings of Variables on the Two Canonical Variates of Their Own Set, for Nominal and Ordinal CANALS Solutions

| | *Nominal* | | *Ordinal* | |
	Dim. 1	*Dim. 2*	*Dim. 1*	*Dim. 2*
1 Reeducation	−.351	.815	−.405	.886
2 Locking up	.049	−.502	−.289	−.302
3 Severe punishment	−.525	.033	−.630	−.159
4 Social work	−.702	−.095	−.719	.106
5 Labor camps	−.557	−.035	−.580	−.313
6 Better employment	−.574	−.193	−.554	.077
7 Watch out in city	−.557	−.437	−.567	−.550
8 Stay home at night	−.630	.018	−.612	−.148
9 Police unreliable	−.735	.538	−.785	.509
10 No help bystanders	.256	.672	−.091	.646
Canonical correlations	.272	.156	.256	.128

who prefer to sit on the fence or are reticent and respondents who do not shrink from expressing outspoken opinions.

5.9 Third CANALS Example: Smoking and Health

5.9.1 Introduction

Data for this example were made available by the Department of Epidemiology of the Institute for Social-Medical Research at the University of Groningen (Van der Lende et al., 1981; Van Pelt et al., 1985). The first set contains variables related to smoking habits:

1. whether the individual smokes
2. how much he or she smokes
3. how long he or she has smoked
4. how long ago the individual had his or her last cigarette

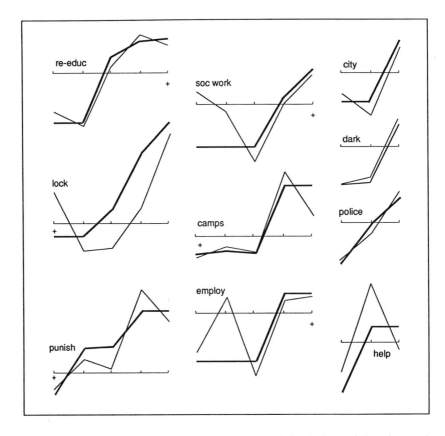

Figure 5.5. Category quantifications for the CANALS solutions of the crime and fear example data. Heavy line indicates single ordinal solution. Thinner line indicates single nominal solution. For the six crime prevention variables, a plus sign marks the "effective" side. The four fear variables run from "fear" (left) to "no fear" (right).

The second set is about health:

5. coughing
6. mucus
7. chest tightness
8. wheezing
9. asthma

TABLE 5.7 Smoking and Health: Loadings on Canonical Variates of Both Sets

| | First Set | | Second Set | |
	Dim. 1	*Dim. 2*	*Dim. 1*	*Dim. 2*
1 Smoke	−.598	.625	−.202	.109
2 How much	−.897	.353	−.303	.062
3 How long	−.737	−.132	−.249	−.024
4 Last time	−.541	.584	−.183	.101
5 Coughing	−.273	.035	−.808	.200
6 Mucus	−.209	−.006	−.619	−.033
7 Tight chest	−.181	−.143	−.535	−.820
8 Wheezing	−.211	.032	−.624	.184
9 Asthma	.003	−.027	.008	−.155
Canonical correlations	.338	.175	.338	.175

This example will also be used in Chapter 6. The present section discusses a CANALS solution in $p = 2$ dimensions, all variables treated as single nominal. There are $n = 2,870$ respondents.

5.9.2 CANALS Results

Category quantifications will not be given. On the whole, they are monotonic, as if the variables were treated as single ordinal. There are a few exceptions for categories with small marginal frequency. Table 5.7 gives canonical loadings and canonical correlations. The latter are not very large.

Interpretation of the first dimension is easy enough. Smoking goes together with unfavorable health symptoms (except asthma). The second dimension shows a small canonical correlation between having a tight chest and how long the person had been a smoker (even if the person was a smoker is long ago). The interpretation might be that the first dimension shows immediate effects of smoking on health, whereas the second dimension is related more to long-term effects of smoking.

6

OVERALS

6.1 Introduction

In classical multivariate analysis the counterpart of OVERALS is generalized canonical analysis (GCA), described in section T.2.4. It is characteristic of GCA that variables can be partitioned into K sets, with K larger than 2. The partitioning should be done a priori, based upon logical and external criteria and not upon the data analysis itself. Later in this chapter an example is presented in which $K = 4$, with variables related to smoking habits, health symptoms, place of residence, and background (sex and age of respondents).

Section T.2.4 explains the general objective of GCA, which is to construct from each set a weighted sum vector, with the weights chosen in such a way that these K sum vectors are as highly as possible intercorrelated. In that case they will also be highly correlated with their first principal component. The latter vector gives the object scores. The exact GCA criterion then becomes to maximize the sum of the squared correlations between object scores and the K weighted sum vectors. This sum of squared correlations is called the *eigenvalue* of the solution. (Sometimes, however, the eigenvalue is identified not with the sum but with the average of the squared correlations.)

Described thus far is the first dimension of a GCA solution. Solutions on subsequent dimensions have the same criterion, but are restricted by the condition that object scores on a subsequent dimension must be uncorrelated with those on all preceding dimensions. This condition implies that object scores still will be given by some principal component of the weighted sum vectors, but not necessarily their first principal component.

6.2 Properties of a Numerical OVERALS Solution

If all variables are treated as numerical, OVERALS comes to the same thing as classical linear GCA. The two methods therefore have some properties in common. The first of these is that the K sets are treated on equal footing. This is the same as in CA or CANALS, where the two sets are treated as symmetrical. In OVERALS, too, no attention is given to some hierarchical order among the sets. Such an order is present when some sets contain "predictor" variables and other sets "predicands." OVERALS will ignore such a difference, when present.

The second thing OVERALS and GCA have in common is that OVERALS can be "unfair" in the sense that a solution depends predominantly upon only some of the K sets, with neglect of other sets. For $K = 4$, for instance, the solution on some dimensions might be determined entirely by strong relations among only three of the sets, whereas the fourth set plays no role.

Third, it may be pointed out at this stage that HOMALS is a variety of OVERALS. In HOMALS, with m categorical variables, each variable is decomposed into a set of k_j binary variables, one for each category. For an illustration it is sufficient to look at an indicator matrix. It thus follows that HOMALS is the special case of OVERALS with m sets of binary variables.

6.3 OVERALS With Categorical Variables

To fix the ideas, suppose that in one of the K sets there are only two categorical variables, each of them with three categories. In linear GCA these categories have a priori quantification. The GCA solution seeks weights for these two variables in such a way that the weighted sum variable of the set will satisfy the GCA criterion. No matter how these weights are chosen, on each dimension there will be only 3×3 category combinations. A weighted sum vector therefore cannot contain more than nine different elements.

In GCA, variables are treated as single in the sense that categories have the same quantification on each dimension of the solution. Suppose now that we take two such dimensions, and that we plot category points on the basis of their coordinates in the two weighted sum vectors. This will result in a plot with 9 category points located on a 3×3 regular lattice in the shape of a parallelogram.

Figure 6.1 shows how such a regular lattice might look. The graph is based upon a numerical example, the details of which we will skip for the moment. Suffice it to say that there are $n = 15$ objects in this fictitious example. The graph also shows their 15 object points. The coordinates of these points are given by the object scores on the two dimensions. Object points will not be located on the 3×3 lattice.

Nevertheless, a good OVERALS solution will imply that the object points must be close to their corresponding lattice points, because otherwise you cannot have high correlation between object scores (based on all K sets) and the weighted sum vector of an individual set. The "ideal" would be that object points coincide with their lattice points. This would imply perfect correlation between object scores and weighted sum vectors, and it would be quite exceptional to find such a result. In fact, this can happen only if there is linear dependence among the weighted sum vectors on each dimension. In other words, variables within one set should measure exactly the same thing as variables in the other sets. One might construe artificial examples in which this happens, but it is extremely unlikely that the same type of result will occur with empirical data.

6.4 OVERALS Criterion

In Figure 6.1 the OVERALS criterion implies, loosely speaking, that object points must be close to their corresponding lattice points. Let us now be more precise. In the figure, the average squared distance between object points and lattice points is called *loss per set*. The smaller the loss, the better the solution. We also could splice loss per set into components for each dimension. Loss per set on the first dimension is based upon

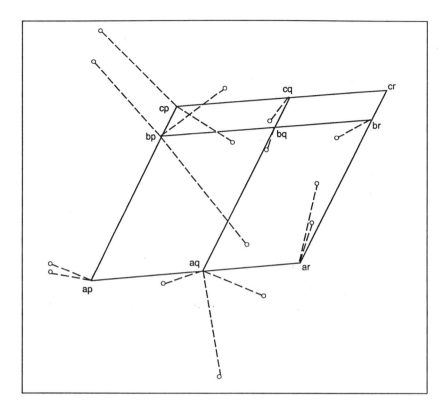

Figure 6.1. For a given set, with two variables (*a, b, c*) and (*p, q, r*), the figure shows the lattice in the shape of a rhombus. Object points are shown as small circles. In the OVERALS solution the criterion is to minimize the average squared distance between object points and their lattice points (dashed lines).

squared horizontal distances, and loss per set on the second dimension is based upon squared vertical distances.

In Figure 6.1 we may also concentrate upon horizontal distances between lattice points and the origin. The average square of these distances represents the spread of lattice points in the first dimension. Its value will be equal to the squared correlation between weighted sum vector and object scores. Averaging over all *K* sets, we obtain the first eigenvalue of the solution. Eigenvalue and total loss per dimension will add up to 1, so that minimizing total loss is equivalent to maximizing the eigenvalue. Similarly, for the second dimension, we should look at vertical distances.

To conclude, if a two-dimensional solution has large sum of the eigen-values, we shall obtain graphs in which the lattice points have large spread with respect to the origin, whereas object points are close to their lattice points. On the other hand, if the two eigenvalues are small, we shall find shrunken lattices, with lattice points close to the origin, and the object points will be far away from their lattice points.

6.5 Nonnumerical OVERALS

Thus far we talked about an OVERALS solution based on a single quantification of the categories of each variable. This applies when all variables are treated as numerical. OVERALS then becomes the same as GCA. Categories then are quantified a priori. The GCA solution is free only with respect to the choice of weights.

Nonlinear OVERALS, on the other hand, selects not only optimal weights, but also optimal quantifications. They may be different from the a priori quantification. In this respect OVERALS has the same options as PRIN-CALS, in that the user may prescribe for each individual variable whether it should be treated as numerical, as single ordinal, as single nominal, or as multiple nominal. The OVERALS solution will be nested as long as all variables are treated as numerical or as multiple nominal. But a solution is not nested once one or more variables are treated as single ordinal or as single nominal. The solution then depends upon the choice of p, the number of dimensions asked for. We saw the same in PRINCALS (Section 4.2.2).

Suppose all variables are treated as single. The OVERALS solution then becomes the same as a GCA solution based upon the optimal quantification of all categories. If by sheer luck the a priori quantification already is optimal, no difference between OVERALS and GCA would be found. But such an event is extremely improbable (the more so since the optimal quantification depends on the choice of p). The only exception, of course, is that all variables are treated as numerical, in which case OVERALS and GCA will always produce identical results.

6.6 Multiple Nominal Treatment of Variables

6.6.1 Broken Lattice

Single treatment of variables results in regular lattices. Figure 6.1 gives an illustration of such a solution. Moreover, this example assumes that there are only two categorical variables in the set. The lattice then has the shape of a parallelogram, and lattice points are the intersections of parallel straight lines. In a situation with more than two variables in a set, and a choice of $p = 3$, we should obtain a three-dimensional lattice. It will look like a parallelopiped, with lattice points at the intersections of parallel planes. With p larger than 3, the lattice will obtain the shape of a "hyperparallelopiped."

However, at this moment the question is what will change in Figure 6.1 if the two variables are treated not as single, but as multiple nominal? The answer is that the lattice no longer will be regular. It will become a *broken lattice*, as illustrated in Figure 6.2. In Figure 6.1 the three lattice points *ap, bp,* and *cp* are located on a straight line, parallel to the line through points *aq, bq,* and *cq* or the points *ar, br,* and *cr.* In Figure 6.2, however, the points *ap, bp,* and *cp* are located on a broken line, of which the line through *aq, bq,* and *cq* is a rigid spatial translation.

The OVERALS criterion remains essentially the same. It is that object points should be close to their corresponding lattice points. But it is obvious that the broken lattice imposes fewer restrictions than the regular lattice. If a regular lattice is replaced by a broken lattice, it will always be possible to obtain shorter distances between object points and lattice points. It follows that when a variable is treated as multiple nominal, loss per set never can be larger than when the same variable is treated as single. A broken lattice can give better fit to the object points than a regular lattice.

6.6.2 Relations With Numerical Solution

Suppose we have an OVERALS solution in p dimensions, and that all variables are treated as single. The latter implies that categories of a variable obtain the same quantification on all p dimensions. Suppose further that this optimal quantification were read in as if it were a priori quantification. Application of numerical GCA on those data then would produce the same result as OVERALS.

On the other hand, suppose that some variables are treated as multiple nominal. Their quantifications then will become different for each dimension. Suppose we take the quantifications according to the first OVERALS dimension and read them in as if they were given a priori. Numerical GCA analysis of these data then will produce the same results on the first dimension as OVERALS, but on subsequent dimensions there will be differences. We might also take quantifications for the second OVERALS dimension and use them as if they were a priori given. Numerical GCA on these data will also turn out with one dimension identical to the second OVERALS dimension. But it is not necessary that this GCA dimension be the second one.

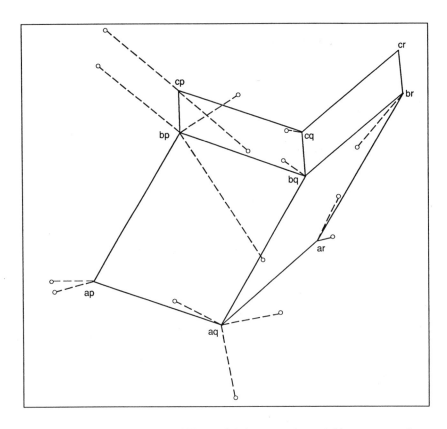

Figure 6.2. Same example as in Figure 6.1, but now the variables are treated as multiple nominal, so that a broken lattice appears.

Formally speaking, the situation is the same as that discussed in Section 2.4 for HOMALS. Given the quantification on some HOMALS dimension, one may apply PCA to the variables thus quantified. It then will be found that one dimension of this PCA solution will be identical to that on the selected HOMALS dimension. Similar results will be obtained if we apply GCA to variables quantified according to one OVERALS solution.

6.7 OVERALS Output

The OVERALS program will, on request, deliver a lot of output. It makes no sense to discuss all the details here. For more information, readers can consult an OVERALS manual or Verdegaal (1986), or Van de Geer (1987). Here only some of the major features of the output are discussed.

OVERALS gives, of course, for each dimension the eigenvalue, equivalent to "total fit" per dimension. Also, the program gives for each set the loss (average squared distance between object points and lattice points). The lattice itself can be regular (if all variables within a set are treated as single), broken (if they are all multiple), or "semibroken" (some variables single, other variables multiple).

OVERALS gives for each variable the discrimination, defined as the variance of the quantified categories. It also gives the category quantifications themselves (in one column for single variables, in p columns for multiple variables), as well as the weights and the loadings. The latter are helpful for arriving at a substantive interpretation of each dimension. Section 6.3.2 discusses the same issue with respect to interpretation of CANALS dimensions. In that section it is also observed that a large canonical correlation does not necessarily imply that a canonical variate explains much of the variance of individual variables within its set. Their loadings on the canonical variate might be surprisingly small. The same can be said about OVERALS. A good OVERALS solution with a large eigenvalue does not imply that the individual variables within a set have high (positive or negative) loadings with respect to their weighted sum variable.

6.8 Multiple Quantification
of Variables Treated as Single

OVERALS output contains another complication, somewhat similar to that found in PRINCALS. It is that one can calculate MC points even for a variable treated as single. They are defined, as usual, as the center of gravity of objects that have the same category combination within a set. In Figure 6.1, for instance, a solution is pictured for two variables treated as single. As a result, the lattice is a regular one. But if we take the centers of gravity of objects with the same category combination, we will obtain MC points on a sort of broken lattice (in the example formed by only eight points, because combination *cr* is empty and therefore has no MC point).

Such a broken lattice may be called an "improper" one, because the OVERALS solution itself is based only upon the regular lattice. Nevertheless, the improper lattice can be used to calculate multiple fit and loss, and these results can be compared with those for single fit and loss per set. If multiple fit appears to be much better than single fit, this is an indication that the OVERALS solution will become much better if the variables are treated as multiple instead of single. But if the difference is small, multiple treatment will produce hardly any gain in fit.

Section 4.3 notes that, for PRINCALS, it would also be possible to take the reverse approach. Given multiple treatment, and therefore a proper broken lattice, one might want to know whether there will be a great increase in loss if the broken lattice is replaced by an improper regular one. However, in PRINCALS such information is not given, and the same is true for OVERALS.

6.9 OVERALS as a "Master Program"

OVERALS can be looked upon as a "master program" in the sense that the programs discussed in the preceding chapters are contained in OVERALS as special cases. The most obvious example is when $K = 2$. OVERALS then becomes essentially the same as CANALS. However, as indicated earlier, there is a subtle difference. CANALS seeks a solution with a maximum for the sum of squared canonical correlations, in p dimensions. OVERALS with $K = 2$, however, maximizes the sum of the absolute values of the canonical correlations. For a solution with just one dimension ($p = 1$) this makes no difference. Nor does it make a difference when all variables are treated as numerical so that the solution is nested. In other situations, where the solution is not nested, a difference may appear. The user should realize that this may be caused by the subtle difference in criteria.

Another special case is that in which each set contains only one variable. OVERALS then becomes the same as PRINCALS. If, in addition, all variables are treated as numerical, OVERALS will be the same as classical PCA. And if all variables are treated as multiple nominal, OVERALS becomes identical with HOMALS. If $K = 2$ and there is only one multiple variable in each of the two sets, OVERALS becomes essentially the same as ANACOR. If there are K sets with only one multiple nominal variable in each set, the OVERALS solution with $p = 1$ will be the same as in the first dimension of PRIMALS.

These properties of OVERALS explain its name. It is an "overall" program. Nevertheless, I do not recommend the use of OVERALS for such special cases where specific programs are available. One reason is that OVERALS will be less efficient in terms of computer time. Another reason is that OVERALS output will not be optimally adapted to special situations. To give just one simple example, OVERALS will maintain such distinctions as "fit per variable" versus "fit per set," even if there is only one variable in each set and the distinction is useless.

6.10 OVERALS Example

6.10.1 Introduction

Before giving the example itself, I would like to state two reservations. They can be explained on the basis of Figures 6.1 and 6.2. Both figures show how loss per set is defined in terms of the average square distance between object points and lattice points. Fit per set is expressed by the average squared distance between lattice point and the origin. The two figures are instructive in that they clarify the meanings of the terms *fit* and *loss*.

However, with real-life data such graphs will not always be attractive. Suppose we have a set with three categorical variables, each with four categories. The lattice will contain $4 \times 4 \times 4 = 64$ points. In a two-dimensional graph such a lattice would be hardly recognizable, even if it were a regular lattice. With a broken lattice the situation becomes even worse.

Further, in real-life examples there are usually many objects. A two-dimensional graph of object points will show a dense cloud of points. It becomes rather hopeless to visualize their distances from lattice points.

6.10.2 OVERALS Example: Smoking and Health

The example is the same as the CANALS example offered in Section 5.9. There we had two sets of variables, one related to smoking habits and the other to health symptoms. For the OVERALS example we add two more sets. The third set contains two background variables, sex and age of respondents. The fourth set contains only one variable, their place of residence. This variable has only two categories, Vlaardingen and Vlagtwedde.

The motivation behind this last variable is that Vlaardingen and Vlagtwedde are two areas that differ greatly in terms of air pollution. Vlaardingen is a relatively small town in the vicinity of Rotterdam. It is surrounded by harbors and industrial plants, oil refineries among them. In Vlaardingen the air is more or less constantly polluted by the refuse of industry and heavy traffic. Vlagtwedde, on the other hand, is a small community in the northeastern part of the Netherlands, located in an agricultural area without heavy industries. In Vlagtwedde the air is relatively pure.

The survey, described in Van der Lende et al. (1981), Van der Burg et al. (1984), and Van Pelt et al. (1985), covered 2,870 respondents: 1,915 from Vlaardingen, 1,510 males, and 847 nonsmokers. The five respiration problems listed in Section 5.9.1 were reported by 277, 207, 644, 680, and 117 respondents, respectively. In the OVERALS example, all variables are treated as single nominal, and a solution in $p = 2$ dimensions was asked for.

6.10.3 Category Quantifications

The two OVERALS dimensions have eigenvalues of .469 and .390, respectively. These eigenvalues are defined here as the average fit per set. About the category quantifications I shall be brief:

1. All five variables about respiration problems obtain higher quantification when the symptom is mentioned, and lower when the symptom is absent.
2. In the third set, the age categories obtain higher quantification to the extent the respondent is older. Sex has positive quantification for males, negative for females.
3. In the fourth set, Vlaardingen has positive quantification, Vlagtwedde negative.
4. In the first set, about smoking habits, quantifications are slightly more complicated. Quantification of the first variable in this set is positive for respondents who smoke or once were smokers, negative for those who never smoked. On the other three variables, nonsmokers obtain zero quantification. Apart from this, quantification increases from light smokers (negative quantification) to heavy smokers (positive).

6.10.4 Component Loadings

Figure 6.3 gives a graph based on the component loadings of all 12 variables. The figure suggests that on the first dimension (vertical direction) variables 3, 4, 7, and 10 are related. This means that having a tight chest occurs more often with older people who have been smokers over a long time and who have had their latest cigarette not so long ago. This relation might be artificial, to some extent. For instance, young respondents cannot have been smokers for very many years, nor can they have had their last cigarette many years ago. What remains is a possible relation between having a tight chest and age.

The second group of related variables is formed by variables 1, 2, 5, 6, 8, 11, and 12. One might say that males (variable 11) tend to be smokers (2), and tend to have respiratory problems (5, 6, 8), especially when they live in Vlaardingen (12).

The question arises about how such a result can be interpreted. Van der Burg et al. (1984) report results of OVERALS analyses performed sepa-

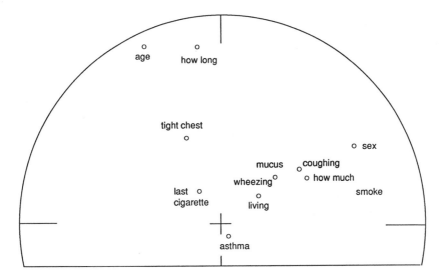

Figure 6.3. Smoking and health. Component loadings of the 12 variables.

rately on males and females. In both groups the relation between tight chest and age was found, but it also appears that within women there is a strong relation between smoking and age (older women do not smoke, younger women do). Within males, variable 12 (place of residence) seems to be irrelevant, but within females this variable makes a lot of difference.

What might be concluded from such results? Vlagtwedde is in many ways a "traditional" community, where not so long ago it was considered unseemly for women to smoke, whereas in Vlaardingen nobody cared. In other words, the difference between Vlagtwedde and Vlaardingen is not only a difference in the amount of air pollution. The two communities also differ in terms of local prejudices, social control, sociographic background, and so on. The message is, again, that results of nonlinear analysis do not provide us with ready-made answers. They may serve mainly as "search models" for future research.

Figure 6.4 pictures component loadings of the four pairs of weighted sums per set. They are labeled y_{ij}, with $i = 1, 2, 3, 4$ for set, and $j = 1, 2$ for dimension. The figure shows that for each of the two dimensions the four corresponding weighted sum vectors form a relatively narrow fan around their axis. In fact, the OVERALS criterion also implies that the larger the eigenvalues, the narrower such fans will be.

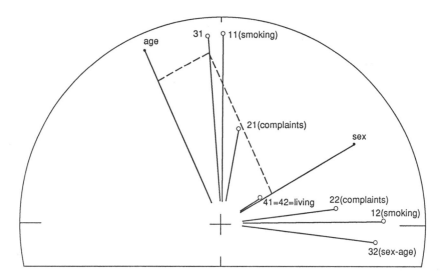

Figure 6.4. Smoking and health. Component loadings of the weighted sum vectors of the four sets.

I conclude with a few technical remarks about Figure 6.4:

1. There is only one variable in the fourth set, so that component loadings of y_{41} and y_{42} are the same as those of variable 12 (place of residence).
2. y_{31} is a weighted sum of age and sex, as indicated by the dotted parallelogram. This weighted sum is located on the arrow for y_{31}, but its coordinates are not the same as those of y_{31} itself. The reason is that y_{31} is standardized, whereas the weighted sum is not.
3. The OVERALS solution is "unfair" in the sense that the fourth set is relatively neglected. This shows merely that place of residence does not play an important role in the relations dominated by the other three sets. From a strict technical point of view it is true that the fourth set becomes more prominent in subsequent OVERALS dimensions. The price is that these dimensions will neglect other sets.

7

Some Conclusions

- 7.1 Stability Analysis
- 7.2 Preview
- 7.3 A Final Consideration

7.1 Stability Analysis

For all GIFI methods, one may ask the important question of how stable the obtained results are. In classical analysis this question is treated in terms of statistical significance. This strategy is usually based upon a number of a priori statistical assumptions, such as the assumption that the variables follow a multinormal distribution. On the basis of such assumptions, a *sampling distribution* can be derived for statistics as the mean or the variance. Confidence intervals can be established, and statistical significance can be evaluated.

In nonlinear research this approach is not very promising, for two reasons. The first is that assumptions about the population distribution are quite arbitrary, hardly more than wild guesses. The second reason is that even if such assumptions about the population hold, it remains extremely difficult to derive sampling distributions for multivariate statistics.

This may explain why an attempt has been made to find other methods for the evaluation of stability. These are known as *Monte Carlo methods*. The idea behind them is quite simple: that computers can be used to generate random quasi-samples. By taking a sufficient number of those random samples, one can obtain a sampling distribution for the statistic one has in mind, and this distribution can be used to establish a confidence interval.

One Monte Carlo technique is known as the *bootstrap*. This approach has become somewhat popular for evaluation of results of nonlinear analysis, especially results of GIFI methods. The basic idea is as follows. Suppose we have a sample with n objects. We then may create a quasi-sample by taking at random n objects from the sample, *with replacement*. The latter means that some of the original n objects may be absent in the quasi-sample, whereas other objects occur more than once. In this way we can form N quasi-samples. If N is sufficiently large, a confidence interval for any statistic can be found. For instance, take the first HOMALS eigenvalue. The bootstrap method will show the distribution of the value of this first eigenvalue for the N quasi-samples. It can be observed that in 95% of the cases the value of the eigenvalue remains between two boundaries. If the lower boundary is larger than some selected numerical value, say .75, one may be confident that the first eigenvalue of the real sample is not smaller than .75.

A confidence interval based on bootstrap does not obey the definition used in classical mathematical statistics, where the boundaries of a confidence interval should be derived from the assumed properties of the population distribution. Nevertheless, comparative studies have shown that in those situations where classical assumptions about the population are met, bootstrap results are in remarkable agreement with classical ones (Efron, 1979; Van der Burg, 1988). Bootstrap therefore seems to be a very reliable method.

7.2 Preview

Prophecies seldom come true. Nevertheless, some future developments of the GIFI system can be predicted. In fact, although they are not yet incorporated in standard GIFI computer programs, some have already been announced in publications. Following are some examples.

Special applications. Some GIFI programs will become better adapted to special applications. One may think of adaptation of CANALS to non-linear discriminant analysis, or of adaptation of OVERALS to multivariate analysis of variance.

REDUNDALS. Classical canonical analysis does not take care of how much of the variance within each set is explained by the canonical variate of the set. An alternative is redundancy analysis. The basic idea behind it is that there are two sets, where the first set contains predictor variables and the second predicands. The question is whether we can find weighted sums of the variables within the first set that will explain as much as possible of the variance of variables in the second set. In redundancy analysis this question is answered for the case where all variables are treated as numerical. It is easy enough to imagine a similar solution for the situation where variables are treated as nominal. We then obtain a nonlinear variety of redundancy analysis. It is obvious that this variety can be called REDUNDALS (Van der Burg, 1988).

PATHALS. Two variables A and C may be correlated because there is a direct interaction between them. It also may happen that A and C are correlated not because of a direct interaction, but because they both depend upon a third variable B. Or it may be that A has a direct influence on B, and B on C, although A has no direct effect on C. On the basis of such ideas the interrelations among a large set of variables might sometimes be explained in terms of a parsimonious model in which it is assumed that there are direct relations only for some pairs of variables, not for all pairs. In classical analysis this approach is called *path analysis*. Its nonlinear counterpart would be PATHALS (Bijleveld, 1989).

Splines. A transformation plot shows the relation between a priori and optimal quantification of categories. When a variable is treated as numerical, such a plot is restricted to a straight line. When the variable is treated as ordinal, the plot must show a monotonic function.

However, a different type of transformation may be required. To give an example, the range of a priori quantifications might be divided into three consecutive trajectories. Within each trajectory one may require different properties of the transformation. One may require, for instance, that in the first trajectory the transformation must be linear, that in the second trajectory it must be quadratic, and in the third trajectory it must be linear again. Moreover, it is required that the three transformations should be connected at their boundaries.

Such solutions can be found by using special adaptations of the indicator matrix. For nominal treatment of a variable, its indicator matrix will consist of binary columns, in which entry 1 is used for "presence" of the column category and entry 0 for "absence." Instead, a more refined coding might be used. Columns of the indicator matrix then are replaced by more sophisticated columns, called *splines*. For a detailed discussion, consult Van Rijckevorsel (1987).

Copies. Suppose we have a categorical variable with $k_j = 4$ categories, and that this variable is treated as multiple nominal on $p = 5$ dimensions. We then obtain five different quantifications of the categories. But these quantified variables cannot have rank larger than $(k_j - 1) = 3$, because four categories cannot be quantified in more than three independent ways.

It then can be shown that the same five solutions for quantification will be found if within its set the variable is included three times (three "copies"), and when it will be required that these three copies must be treated as single nominal. In other words, each of the five "multiple" quantifications will be a weighted sum of the three "single" quantifications of the copies.

Thus far the lemma above seems to reflect no more than an algebraic triviality, but it becomes more interesting if we take less than $(k_j - 1)$ copies. In the example above, suppose we take only two copies, both to be treated as single nominal. It follows that on the $p = 5$ dimensions the quantifications will appear as weighted sums of the two single solutions. They therefore will have rank 2.

What is the advantage? As another example, take a variable "age" with $k_j = 10$ categories, for 10 successive age intervals. With $p = 5$, multiple treatment of this variable will probably result in different quantifications that have rank 5. But if we take only two copies, both treated as single, the five quantifications are limited to rank 2. Moreover, it may happen that the single quantifications of the two copies have an easy interpretation. For instance, in one of the copies the transformation is almost linear, and in the other one almost quadratic. It then follows that quantifications on the five dimensions will be weighted sums of a linear and a quadratic function of age, so that their interpretation remains easy enough.

On the other hand, the possibility of using copies increases the *embarras du choix*. How many copies should be taken? And how should they be treated? For instance, should two copies be taken, both treated as single nominal? Or one of them as single ordinal? Or both single ordinal?

It may be a good suggestion to start with only one copy, treated as multiple nominal. Suppose it appears that the p quantifications have rank close to 2. The analysis then may be repeated with two copies, both treated as single nominal, to the effect that the p quantifications will have exactly rank 2. Moreover, if it looks promising, one of the two copies might be treated as numerical, or as single ordinal.

We have seen in Chapter 4 of this volume that PRINCALS has no option for treating a variable as multiple ordinal. It now appears that this limitation can be circumvented by taking two (or perhaps more) copies to be treated as single ordinal. The result might show that one of the ordinal transformations is flattened toward the end (like a logarithmic function), whereas the other one is flattened at the beginning (like an exponential function). If the variable is about age, the first quantification discriminates primarily between young respondents and the second one between older respondents. (For further discussion of copies, see De Leeuw, 1984b.)

Distances. Multivariate analysis and multidimensional scaling are two topics with their own historical developments, with surprisingly little interaction between the two. The focus of multivariate analysis always has been on matrix algebra, whereas in scaling the primary theoretical interest is in the concept of distances. There is, of course, no essential antithesis between these two approaches. Multivariate algebra always can be reformulated in geometric terms, and vice versa. That multivariate analysis and multidimensional scaling followed their own ways, with relatively little cross-fertilization between them, seems to have a historical explanation only. I expect that the two approaches will grow closer together (for an example, see Meulman, 1986).

7.3 A Final Consideration

To conclude this book I would like to discuss just one special feature of the GIFI system: its many options. For example, there is the question of whether variables should be treated as one set (as in HOMALS or PRIN-CALS) or should be partitioned into two or more sets (as in CANALS or OVERALS). There is the decision to treat some variables as passive. And should the active variables be treated as numerical, ordinal, single, or multiple nominal? Should copies be used in order to restrict the rank of different quantifications? How many dimensions should be prescribed by the user (particularly relevant if dimensions are not nested)? How should missing data be handled? Further, how should they be defined?

There is no foolproof set of rules that instructs the user which of the options are "best." Coombs (1964), in his book *A Theory of Data,* makes a profound distinction between *observations* and *data.* In the present book this distinction is not maintained; instead, the term *data* is used quite loosely. According to Coombs, however, what one has registered as the outcome of one's research defines what the observations are. When one starts analyzing them, the first step is to transform observations into data.

The idea is quite simple. If a researcher decides to treat a variable as ordinal, the data are not given by some a priori quantification of categories, but only by their rank order. Or if a researcher decides to treat a category as standing for "passive missing data," compared with handling objects within that category as belonging to the same subgroup, he or she creates different data. Observations are "given" and the researcher can do nothing about them, but it is the researcher who decides what the data are.

Such decisions depend on what questions the researcher wants to answer. A researcher may start with an analysis that includes all observed variables, but results of the analysis may depend upon variables in which he or she is not primarily interested. The researcher thus realizes that he or she implicitly made a sort of distinction between *target variables* and *incidental variables.* Keeping this distinction in mind, the researcher should reformulate the initial questions. This could imply that the incidental variables should be treated as passive and only target variables taken as active. This creates a new set of data. The new set of data also will answer a new set of questions. Similarly, the researcher may become aware that the questions require him or her to make a distinction between two sets of active variables. This implies that the researcher creates data on which CANALS can be applied.

In this way, analyzing observations becomes a *search process* during which the researcher learns, step by step, what the question are that he or she really wants answered. The implication is that observations often need to be analyzed in more than one way, sometimes in many ways. If there are discrepancies between results of such different analyses, the researcher must try to unravel why they occur and to discover what different questions are answered by them. Perhaps the researcher discovers that the original questions were too naive, and must be discarded altogether. Perhaps the original questions were too hazy and need sharpening before meaningful answers can be expected.

Seen in this light, the *embarras du choix* implied in the large number of GIFI options is not so much a nuisance as it is a tool for research.

Appendix:
Dutch Political Parties

The Dutch Parliament has two chambers. The 150 members of the Second Chamber are chosen in direct elections, and are represented in proportion to the votes received by their parties. There is no threshold, and as a consequence small parties may succeed in obtaining one or two seats, sometimes for short periods only. The 75 members of the First Chamber are elected in a more complicated way, but on the whole the political parties are represented in proportion to the popular vote.

The Second Chamber is directly involved in legislation. The function of the First Chamber is mainly to see to it that proposed legislation is in agreement with the Constitution.

In this book, references are made to 16 different parties. Below are admittedly impressionistic and oversimplified descriptions of these. My only objective here is that a reader who is unfamiliar with the Dutch political system may have a better understanding of the examples in this book.

First, there are four relatively large parties:

1. Partij van de Arbeid (PvdA): 28% of the vote (averaged over the years between 1963 and 1982); a social-democratic party, somewhat comparable to Labor in the United Kingdom or SPD in Germany

2. Christen Democratisch Appel (CDA): 31% of the vote; a denominational party, although not very outspoken in this sense; formed in 1976 as a merger of three former denominational parties—Anti Revolutionaire Partij (AR), 9%, a Protestant party; Katholieke Volkspartij (KVP), 25%, a Roman Catholic party; Christelijk Historische Unie (CHU), 7%, a Protestant party with a somewhat "liberal" tinge

3. Volkspartij voor Vrijheid en Democratie (VVD): 15% of the vote; called a "liberal" party, but generally perceived as a "conservative" party

4. Democraten 1966 (D'66): 6% of the vote; founded in 1966, as a sort of pragmatic and rejuvenating alternative for the "old" parties; electoral success has been up and down, but in recent elections did quite well

Next are four parties that could be collected under the debatable name of "small right." Their percentage of the vote tends to be below 3%.

1. Centrum Partij (CP): a nondenominational party, considered by others to be somewhat undemocratic, or even "racist"

2. Staatkundig Gereformeerde Partij (SGP): based upon what some people would call "puritan Protestantism"; supported mainly by members of the dogmatic Reformed church

3. Gereformeerd Politiek Verbond (GPV): somewhat like SGP, but less dogmatic

4. Reformatorische Politieke Federatie (RPF): somewhat like SGP or GPV

Supporters of the three political parties last mentioned will object vehemently to the superficial descriptions above in which these three parties are sketched as "somewhat similar." In fact, the three parties differ quite a lot with respect to their interpretations of a number of theological principles.

There are also four parties that might be collected under the label "small left." Their percentage of the vote is below 3%.

1. Communistische Partij Nederland (CPN): a traditional communist party, at least in the period to which the examples refer

2. Partij Politieke Radicalen (PPR): founded about 1970 by dissidents from the denominational parties (KVP in particular), but gradually moved away from this background and became more and more an ideological leftist party

3. Pacifistisch Socialistische Partij (PSP): emphasizes pacifism and leftist ideological principles

4. Evangelische Volkspartij (EVP): represented in the First Chamber for the first time in 1982; supporters are mainly Protestants, with a political inclination toward the left

Finally, there is the party called Democratisch-socialisten 1970 (DS'70). This party was founded in 1970 by dissidents from PvdA in protest against what they interpreted as a dangerous movement within PvdA to move toward an extreme leftist ideology. DS'70 obtained seats in Parliament in the period 1971-1981. Afterward, the party was dissolved, probably because it had little popular support in terms of votes. Another reason might be that representatives of this party in Parliament seemed to take different stands with respect to some political issues.

Recommended Reading

General Background

Benzécri, J. P., et al. (1973). *Analyse des données.* Paris: Dunod.
de Leeuw, J. (1984a). *Canonical analysis of categorical data.* Leiden, Netherlands: DSWO Press.
de Leeuw, J. (1984b). The GIFI-system of non-linear multivariate analysis. In E. Diday, M. Jambu, L. Lebart, J. P. Pages, & R. Tomassone (Eds.), *Data analysis and informatics III* (pp. 415-424). Amsterdam: North Holland.
GIFI, A. (1990). *Nonlinear multivariate analysis.* New York: John Wiley.
Greenacre, M. J. (1984). *Theory and applications of correspondence analysis.* London: Academic Press.
Nishisato, S. (1980). *Analysis of categorical data: Dual scaling and its applications.* Toronto: University of Toronto Press.
Van de Geer, J. P. (1986). *Introduction to linear multivariate data analysis.* Leiden, Netherlands: DSWO Press.

PRIMALS

Van de Geer, J. P., & Meulman, J. J. (1985). *PRIMALS user's guide.* Leiden, Netherlands: University of Leiden, Department of Data Theory.

116

HOMALS

Guttman, L. (1941). The quantification of a class of attributes: A theory and method of scale construction. In P. Horst (Ed.), *The prediction of personal adjustment* (pp. 319-348). New York: Social Science Research Council.

Lord, F. M. (1958). Some relations between Guttman's principal components of scale analysis and other psychometric theory. *Psychometrika, 23,* 291-296.

Meulman, J. J. (1982). *Homogeneity analysis of incomplete data.* Leiden, Netherlands: DSWO Press.

Van de Geer, J. P. (1985). *HOMALS user's guide.* Leiden, Netherlands: University of Leiden, Department of Data Theory.

ANACOR

GIFI, A. (1985). *ANACOR user's guide.* Leiden, Netherlands: University of Leiden, Department of Data Theory.

PRINCALS

GIFI, A. (1983). *PRINCALS user's guide.* Leiden, Netherlands: University of Leiden, Department of Data Theory.

CANALS

Van der Burg, E. (1983). *CANALS user's guide.* Leiden, Netherlands: University of Leiden, Department of Data Theory.

Van der Burg, E. (1988). *Nonlinear canonical correlation and some related techniques.* Leiden, Netherlands: DSWO Press.

OVERALS

Van de Geer, J. P. (1987). *Algebra and geometry of OVERALS.* Leiden, Netherlands: University of Leiden, Department of Data Theory.

Van der Burg, E. (1988). *Nonlinear canonical correlation and some related techniques.* Leiden, Netherlands: DSWO Press.

Verdegaal, R. (1986). *OVERALS user's guide.* Leiden, Netherlands: University of Leiden, Department of Data Theory.

References

Benzécri, J. P., et al. (1973). *Analyse des données*. Paris: Dunod.

Bijleveld, C. C. J .H. (1989). *Exploratory linear dynamic systems analysis*. Leiden, Netherlands: DSWO Press.

CBS. (1978). *Statistiek winkeldiefstal 1977*. The Hague: Staatsuitgeverij.

CBS. (1979). *Statistiek winkeldiefstal 1978*. The Hague: Staatsuitgeverij.

CBS. (1983). *Statistisch zakboek 1983*. The Hague: Staatsuitgeverij.

CBS. (1987). *Statistiek der verkiezingen 1986. Tweede Kamer der Staten Generaal 21 mei*. The Hague: Staatsuitgeverij.

Coombs, C. (1964). *A theory of data*. New York: John Wiley.

Cozijn, C., & Van Dijk, J. J. M. (1976). *Onlustgevoelens in Nederland*. The Hague: WODC.

Daalder, H., & Rusk, J. B. (1972). Perceptions of party in the Dutch Parliament. In S. C. Patterson & J. C. Wahlke (Eds.), *Comparative legislative behavior: Frontiers of research* (pp. 143-198). New York: John Wiley.

de Leeuw, J. (1984a). *Canonical analysis of categorical data*. Leiden, Netherlands: DSWO Press.

de Leeuw, J. (1984b). The GIFI-system of non-linear multivariate analysis. In E. Diday, M. Jambu, L. Lebart, J. P. Pages, & R. Tomassone (Eds.), *Data analysis and informatics III* (pp. 415-424). Amsterdam: North Holland.

de Leeuw, J., & Van Rijckevorsel, J. (1988). Beyond homogeneity analysis. In J. L. A. Van Rijckevorsel & J. de Leeuw (Eds.), *Component and Correspondence Analysis*. Chichester, UK: John Wiley.

Diekstra, R. F. W., & Kerkhof, A. J. F. M. (1982). *Suicide pogingen in de bevolking*. Leiden, Netherlands: University of Leiden, Vakgroep Klinische Psychologie.

Efron, B. (1979). Bootstrap methods: Another look at the jackknife. *Annals of Statistics, 7,* 1-26.

GIFI, A. (1983). *PRINCALS user's guide.* Leiden, Netherlands: University of Leiden, Department of Data Theory.

GIFI, A. (1990). *Nonlinear multivariate analysis.* New York: John Wiley.

Greenacre, M. J. (1984). *Theory and applications of correspondence analysis.* London: Academic Press.

Guttman, L. (1968). A general nonmetric technique for finding the smallest space for a configuration of points. *Psychometrika, 33,* 469-506.

Hudson, F. R. (1968). *The La Thène cemetry at Münsingen-Rain: Catalogue and relative chronology.* Bern: Stämpli.

Israëls, A. Z. (1987). *Eigenvalue techniques for qualitative data.* Leiden, Netherlands: DSWO Press.

Lingoes, J. C. (1968). The multivariate analysis of qualitative data. *Multivariate Behavioral Research, 3,* 61-94.

Maung, K. (1941). Measurement of association in a contingency table with special reference to the pigmentation of hair and eye colours of Scottish school children. *Annals of Eugenetics, 11,* 189-223.

Meulman, J. J. (1986). *A distance approach to non-linear multivariate analysis.* Leiden, Netherlands: DSWO Press.

Speckmann, J. D. (1965). *Marriage and kinship among the East Indians in Suriname.* Assen, Netherlands: Van Gorcum.

Speyer, N., & Diekstra, R. F. W. (1980). *Hulp bij zelfdoding, een studie rond de zelf gewilde dood.* Deventer, Netherlands: Van Loghum Slaterus.

SPSS Categories. (1990). Chicago, IL: SPSS Inc.

Van de Geer, J. P. (1985). *HOMALS user's guide.* Leiden, Netherlands: University of Leiden, Department of Data Theory.

Van de Geer, J. P. (1986). *Introduction to linear multivariate data analysis.* Leiden, Netherlands: DSWO Press.

Van de Geer, J. P. (1987). *Algebra and geometry of OVERALS.* Leiden, Netherlands: University of Leiden, Department of Data Theory.

Van der Burg, E. (1983). *CANALS user's guide.* Leiden, Netherlands: University of Leiden, Department of Data Theory.

Van der Burg, E. (1988). *Nonlinear canonical correlation and some related techniques.* Leiden, Netherlands: DSWO Press.

Van der Burg, E., et al. (1984). *Non-linear canonical correlation with m sets of variables.* Leiden, Netherlands: University of Leiden, Department of Data Theory.

Van der Lende, R., et al. (1981). Decreases in VC and FEV with time: Indicators for effects of smoking and air pollution. *Bulletin Européen de Psychologie Resparitoire, 17,* 775-792.

Van Pelt, W., et al. (1985). Analysis of maximum expiratory flow volume curves using canonical correlation analysis. *Methods of Information in Medicine, 24,* 91-100.

Van Rijckevorsel, J. (1987). *The application of fuzzy coding and horseshoes in multiple correspondence analysis.* Leiden, Netherlands: DSWO Press.

Verdegaal, R. (1986). *OVERALS user's guide.* Leiden, Netherlands: Department of Data Theory, University of Leiden.

Author Index

Subject index

About the Author

John P. Van de Geer completed his study in psychology at the University of Leiden, Netherlands, in 1953. He received his Ph.D. in 1957, and served as Professor of Experimental Psychology and Psychological Statistics at the University of Leiden from 1960 to 1970. From 1958 to 1968 he was also an adviser to the Institute for Perception RVO/TVO, Soesterberg, Netherlands. Thereafter, he was Professor of Data Theory at the University of Leiden until his retirement in 1988. In 1989 he received the Royal/Shell Award for his contribution to the social sciences. His main research interests have been in the areas of experimental psychology and research methodology, with a gradual shift from the first field toward the second. He has published many papers on perception (visual and auditory) and on data analysis. His books include a Dutch volume on multivariate analysis published in 1967, followed by *Introduction to Multivariate Analysis for the Social Sciences* (1971), *Introduction to Linear Multivariate Data Analysis* (2 volumes; 1986), and a Dutch book on analysis of categorical data, of which the present two-volume set is a translation.